# The Seven Secrets of How to Think Like a Rocket Scientist

# The Seven Secrets of How to Think Like a Rocket Scientist

Jim Longuski, Ph.D.

**Copernicus Books**
*An Imprint of Springer Science+Business Media*

Published in the United States by Copernicus Books,
an imprint of Springer Science+Business Media.

Copernicus Books
Springer Science+Business Media
233 Spring Street
New York, NY 10013
www.springer.com

*Library of Congress Control Number:*
2006922755

Manufactured in the United States of America.
Printed on acid-free paper.

9 8 7 6 5 4 3 2

ISBN-10: 0-387-30876-8
ISBN-13: 978-0-387-30876-0

# Acknowledgments

I'd like to thank all those who contributed to this book by their positive support, their helpful suggestions, and their sharp eyes for typos. Among these are my friends and colleagues Dr. James R. Wertz and Professor Tasos Lyrintzis; Purdue graduate students Janelle Boys, K. Joseph Chen, Karl Garman, Damon Landau, Kristin Medlock, Masataka Okutsu, Tracey Smith, Christoph Wagner, and Chit Hong "Hippo" Yam; my brother, Joseph A. Longuski, and my mother, Jeanette T. Longuski, and Ronit Binder, Dr. Michael Jokic, Dr. T. Troy McConaghy, Elma Witty, and Wendy Witty.

I thank my secretary, Karen L. Johnson, for her work in typing the corrections and for several thoughtful improvements. I thank doctoral candidate Mr. Masataka Okutsu for his delightful interior and cover illustrations. I also thank Dr. Harry (J.J.) Blom and Mr. Chris Coughlin (senior editor and assistant editor for astronomy & astrophysics at Springer) and Mr. Michael Koy (senior production editor at Springer) for promoting the publication of my book and for their kindness and patience throughout the process. Thanks also to freelance editor Paul Farrell for his advice and counsel, and to graphics consultant Jordan Rosenblum for making the cover and interior designs sparkle.

I hasten to add that the individuals named here (and those who have written encomia for the book) do not necessarily agree with all the opinions expressed by the author.

Above all, I thank my wife and best friend, Holly C. Longuski, who has given me the greatest encouragement and support.

# Contents

# Introduction

This is a book for the armchair thinker. There are no equations, no syllogisms, and no exercises with the solutions at the back of the book.

It is written not for rocket scientists (although they might enjoy it, too) but for the non–rocket scientist.

Before I wrote this book, I asked a number of people what they hoped to find in a book about how to think like a rocket scientist. "Do you want to know what rocket scientists actually think about and have it translated into ordinary language?" I asked, and everyone said, "No."

"Then would you prefer to know the methods that rocket scientists use—not the content—expressed in a way that you could apply to your everyday life?"

And then everyone said, "Yes."

The book you are holding does just that. (Mostly.)

Let me tell you the first secret about rocket scientists. They are not in it for the money. They are in it for the fun. They are the biggest dreamers on Earth because they dream on a cosmic scale. And they love sci-fi books and movies. Sometimes, the dopier the movie, the better they like it.

That's why I start Part I with "Dream." Dreaming about space travel is what makes rocket scientists tick. I end with Part VII, "Do," because the best part about rocket science is when those dreams come true. I give seven secrets of how to think like a rocket scientist as active verbs: "Dream," "Judge," "Ask," "Check," "Simplify," "Optimize," and "Do."

I talk about how we can all use some of the thinking techniques that rocket scientists learned from the extraordinary challenges of space exploration. This doesn't mean that rocket scientists are all geniuses or that they never make mistakes. They have been humbled often enough by catastrophic explosions, destruction of billion-dollar satellites, and loss of life.

A great deal of effort is put into avoiding mistakes because mistakes are so costly. But some of the greatest lessons came from the worst failures.

The best known rule of thumb in the space business is Murphy's law: "If something can go wrong—it will!" Space history revolves around the struggle of beating Murphy's law.

In this book, I have written several short chapters about each of the seven secrets of how to think like a rocket scientist. I illustrate the principles with anecdotes, quotations and biographical sketches of famous scientists, ideas from sci-fi, personal stories and insights, and occasionally a bit of space history. At the back of the book, I give, not the solutions to brain teasers, but instead a list of imagination builders: my list of the greatest science fiction movies of the twentieth century. (The jury is still out on the twenty-first century.) I also provide a Recommended Reading and Bibliography list.

In the course of writing this book, I found it necessary to distinguish between "two NASAs": the NASA that put men on the moon and the NASA that built the space shuttle. From the original NASA, we can learn how to think like rocket scientists. Unfortunately, the latter NASA provides examples of how not to think like rocket scientists. Because my goal is to provide you, the reader, with thinking techniques distilled from the space program, I draw from the historical record—good and bad. I hasten to add that my occasional criticism of NASA as an institution in no way diminishes my admiration for its highly qualified scientists, engineers, technicians, and staff—some of the best talent in the world—who yearn for far greater challenges (and the requisite funding) to explore space.

I hope you enjoy this little collection of ideas and find some of the techniques useful.

# PART I

# Dream

"His adventure began with a dream . . . Robert Goddard had a waking dream about flying farther than anyone ever had, to other worlds in the sky."

David A. Clary
*Rocket Man*

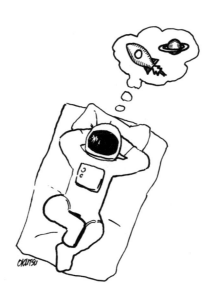

# 1

## Imagine It

If you could not fail, what would you attempt?

Forget about your fears, the facts, looking silly or stupid—and test your ability to dream.

Albert Einstein said that imagination is more important than knowledge. Why would he say something so contrary to his pursuit of scientific truth? To free his imagination. To suspend his fear of being wrong—for a while—and to dream how the universe might be.

What would you dream?

Rocket scientists have their answer. Rocket scientists love science fiction novels and movies: stories about traveling to Mars, Jupiter, Alpha Centauri, the Andromeda Galaxy; about contact with alien beings, many-tentacled monsters, conscious robots, and giant ants (or spiders or locusts or gorillas). Their favorite books are not literature. Their favorite films are the exemplars of B-grade movies. So what does this demonstrate about rocket scientists?

They aren't afraid of looking silly.

How can a rocket scientist who has remotely piloted a deep space probe to the outer fringes of the solar system enjoy the 1950 film *Destination Moon*, which tenders a juvenile plot, serves up wooden dialogue, and features cheesy special effects?

Let's take a closer look at a group of such rocket scientists who worked for a prestigious government laboratory. On a regular basis, they would meet for a "Sci-Fi Film Festival" in which they'd watch 1950s videos. They'd watch such classics as *The Day the Earth Stood Still* and *Forbidden Planet* and such crap as *Plan 9 From Outer Space* and *I Married A Monster From Outer Space*. They memorized lines like "Gort, Klaatu barada nikto!" (what to say to the robot to stop him from vaporizing you) and "The fool—to think that his ape-brain could contain the secrets of the Krell!" (what Dr. Morbius said to the rescue ship's doctor who took the IQ boost). They'd laugh at the bad navigation in *Rocketship X-M* where the spacecraft "accidentally" goes to Mars instead of the moon.

But they loved these films.

They were like children who want to hear the same fairy tale over and over again. These were the fairy tales of the rocket scientists; their unfettered hearts seeking contact with outer space. Their logic turned off (their humor kept on)—their dreams turned on.

Imagination wasn't silly to them.

# 2

## Work on the Big Picture

In *Advice to Rocket Scientists*, I talk about two bricklayers who are asked by a young boy what they are doing. The first bricklayer is annoyed at the question and says, "Can't you see? I'm laying bricks." The second says with a gleam in his eye, "I'm building a cathedral!"

The first bricklayer was a little-picture person. All he could see was the tedious job of laying one brick at a time. The second bricklayer was a big-picture person. He envisioned a beautiful cathedral in all its glory and he reveled in his task to help create it.

Find your big picture and it will give your task perspective and joy. The big picture focuses your mind and subconscious on a larger purpose. It gives meaning to all the little tasks you must tend to in order to achieve your goal.

The Chinese philosopher Lao Tse said that "a journey of a thousand miles begins with a single step." If we could ask him where he was going, he'd probably describe a distant land of great enchantment. If we could ask him how he expected to get there, he'd demonstrate silently by taking another step. Keep your big picture in mind when solving your problems. The big picture will help you take the next step—it will give you direction.

Albert Einstein was always looking for the most general theory to explain how the universe operates. He explained the mysterious constancy of the speed of light by his special theory of relativity. In this case "special" meant restricted. Later, he removed the restriction and came up with his general theory of relativity, which explained how gravity works. Einstein then tackled the most difficult problem of all: to develop a unified field theory to explain not only gravity but all the forces in the universe.

Einstein spent only a few years developing his special theory, a decade for his general theory, and the rest of his life searching for a unified theory. Einstein was a big-picture person. He was not interested in how a particular atom vibrates—he wanted to understand the entire universe. His big picture gave him direction throughout his scientific life. Not all scientists think Einstein was right. But today, many are working on the "theory of everything." Einstein's big picture continues to inspire new generations.

# 3

## Aim High

Rocket scientists aim high. They reach for the moon and beyond. Their dreams are gigantic in scale. They may not always achieve their goals, but they know that you never hit a target that you don't aim at. (As hockey great Wayne Gretsky said, "You miss 100 percent of the shots you don't take.") Sometimes their dreams come true, but even when they don't, the achievements of rocket scientists are great.

Ernest Shackleton, the polar explorer, aimed high. Maybe we should say he aimed low, because his target was the South Pole. In 1902, he traveled with Robert F. Scott to within 460 miles of the pole. In 1908, he commanded his own expedition but was forced back after falling short of the pole by 97 miles. To have gone on to reach his goal would have meant certain death to his crew. Though Shackleton was criticized by some, he considered the safety of his men to be of far greater importance than his stated mission. Scott, who was rigorously trained in the British navy, was of the school that some loss of life was inevitable. Similar arguments have been made in defense of the space shuttle, but as we shall see later, there are better, safer ways to explore space.

On December 14, 1911, Shackleton's dream was dashed when Roald Amundson of Norway reached the South Pole. One month later, Scott and his party reached the pole but died on their return trip. In the next few years, Shackleton, undaunted by the success of Amundson, planned a daring adventure: the first transcontinental expedition of Antarctica.

In the attempt he made his greatest failure. He lost his ship but saved every member of his crew in a dramatic two-year misadventure (told in a terrific book by Margot Morrel and Stephanie Capparell: *Shackleton's Way*).

Shackleton failed in nearly every mission he launched, and yet he is considered today to be the greatest of the Antarctic explorers. He aimed high, but he changed his plans to fit the circumstances— he didn't believe in Pyrrhic victories, and he didn't lose a man in his command.

He is a shining example of how we should approach human exploration of Mars and beyond.

Not long ago, a crater was discovered on the moon that circumscribes the lunar South Pole. It was named in Shackleton's honor. Someday, astronauts may explore the depths of Shackleton Crater— a region of eternal darkness—to search for a substance more precious than gold: ice.

# 4

## BS!

Unfortunately, when we talk about creativity, about generating new ideas, and about solving difficult problems, most people become stiff and formal. You may think that creativity is an activity left to the erudite—the well-mannered professor, the dignified inventor in a lab coat, the rocket scientist (our hero). There is a strong tendency to become judgmental and critical, to get serious, and to not be creative at all.

This is what happens when you ask people to get creative. Think about all the boring stories that have been written about "How I spent my summer vacation."

What's wrong with this picture? It's that people constrain themselves, they look for answers that seem acceptable to whomever they are trying to please—they try to stay safely inside the box.

They are afraid to offend, to make a mistake, to appear irreverent or nonchalant, to look silly. And thus, you just can't be creative when someone tells you to be creative.

But, on the other hand, everyone knows how to BS. (It's a good word—as Henry Fonda told us in *On Golden Pond*.) BS is making stuff up, telling stories, trying to amuse, and is definitely irreverent. BS knows no decorum, no bound, no fear, and no respect. We all do it. BS is fun, BS is playful, BS is creativity without constraints.

Do rocket scientists BS? They sure do! They love to do it and they love to hear it. Why do they like sci-fi so much?

So is that all they do—just make it up as they go along? Is that all you need to know? Of course not—you should know better than that! There is a time for BS and a time for separating the good ideas from the bad. (We'll discuss this in Part II: Judge.)

You can't get away with BSing your way through with just any BS. It's got to be good BS. You've got to be able to sniff out the wheat from the chaff.

# 5

## Brainstorm

The first step to knowledge—to finding the answer—is to eliminate what isn't true. Thomas Edison, during his struggle to create the incandescent light bulb, performed thousands of unsuccessful trials. When reporters asked him what he thought about his lack of progress, he replied: "I haven't failed. I've just found ten thousand ways that won't work!"

Sir Arthur Conan Doyle gave us many lessons about thinking through his brilliant fictional detective, Sherlock Holmes, who explains his methods to his sidekick, Dr. Watson. On one occasion, Holmes tells Dr. Watson, "When you have eliminated the impossible, whatever remains, however improbable, must be the truth." By the way, this is not a bad strategy when taking multiple-choice tests (such as the SATs, GREs, and IQ tests) where you may not be sure of the right answer but can appear a lot smarter by eliminating the answers you know are wrong.

We get another glimpse into the mind of genius when, early on in an investigation, Dr. Watson asks Holmes to reveal his current hunch about a crime. "I have devised seven separate explanations, each of which would cover the facts as far as we know them. But which of these is correct can only be determined by . . . fresh information. . . ."

We see the fundamental concepts of trial and error, of hypothetical solution generation and elimination, of brainstorming, and of judging. We will discuss judging later. For now we will concentrate on the process of entertaining many different solutions simultaneously.

Brainstorming consists of making a long list of possibilities. The goal is to create as many ideas (the good, the bad, and the ugly) as you can, to make your list as long as possible. (Here's a situation where length really does matter.)

No idea, no matter how absurd, stupid, ridiculous, or silly, should be discarded. Absolutely no judgment should be made at this stage. Turn your judging mind (your logic center) off. Make it a kitchen-sink argument: throw everything at your problem (but the kitchen sink). You can do this exercise alone or with a group, but it's more fun with a group. Too many cooks will not spoil the broth.

Give your creativity (and everyone else's) free rein. (And remember—BS, which could also stand for brainstorming, works!) Don't take the process too seriously. Don't be afraid to play with ideas. The time to criticize will come later. You may be searching for a needle in a haystack, but you first must build the haystack.

To land a man on the moon within the decade, as President Kennedy directed, rocket scientists had a preconceived idea. We would build a rocket that would launch from the surface of Earth and would land all three men directly on the surface of the moon. After planting their flag, leaving their footprints in the dust, and collecting rock samples, the three men would blast off from the lunar surface and return directly to Earth.

But that is not how it was accomplished.

Original calculations demonstrated that a super rocket, dubbed the *Nova*, would have to be constructed. It would be 500 feet tall (as tall as a 50-story building) and would weigh 12 million pounds. It seemed impossible to everyone but Wernher von Braun who dreamt of this super rocket.

But then a NASA Langley engineer, Dr. John C. Houbolt (which rhymes with cobalt) proposed a different approach: lunar orbital rendezvous. Instead of landing all three men on the moon along with the return rocket, Houbolt suggested parking the return vehicle in orbit around the moon, piloted by one of the astronauts. A much smaller vehicle, the lunar module, would take two astronauts down to the lunar surface and later back up to the lunar parking orbit. The lunar module would be incapable of returning to Earth on its own power. It would have to rendezvous with the mother ship (the command service module), which had the propellant to send the three astronauts home.

When the concept of lunar rendezvous was suggested, it was considered crazy and dangerous. At first, NASA's famous spacecraft designer, Max Faget, was a bitter opponent and told Houbolt, "Your figures lie!" If the men were unable to catch up with the mother ship, they would die in lunar orbit. Only the astronaut in the command module could return alive.

The idea was the result of brainstorming—thinking outside the box. And it changed everything. Rocket scientists would have to develop a technique to join two spacecraft in space. The concept was tested during the *Gemini* program in the relative safety of Earth orbit. After some harrowing, nearly fatal missions, rendezvous and docking in space was successfully demonstrated.

Lunar rendezvous eliminated the need to build the rocket of von Braun's dreams (the *Nova*) and allowed men to reach the moon with the much smaller *Saturn V* rocket. The *Saturn V* was still gigantic at 365 feet tall (36 stories) and weighing 7 million pounds. This little brother of the *Nova* got America to the moon in only eight years, easily fulfilling President Kennedy's challenge. But without the brainchild of John Houbolt, it would never have happened. Moments after Mission Control confirmed that Armstrong and Aldrin had safely landed on the moon, Werner von Braun turned to Houbolt and said, "John, it worked beautifully."

# 6

## Create Desire

"You've got to have *ganas*—desire," Mr. Escalante told his high school class of Mexican-American students in East Los Angeles. "Desire to know."

Jaime Escalante's story is immortalized in the inspiring film *Stand and Deliver*. In the movie, Mr. Escalante has left a lucrative job as a computer programmer in the aerospace business to take on a low-paying teaching position in East L.A. where he faces off with the class's cholo gang leader. Undaunted, Mr. Escalante tells his students that "you have mathematics in your blood." He tells them that without an education they will end up pumping gas for a living. To convince his students that they can be successes, Escalante invites an F-16 pilot to lecture on the importance of mathematics in flying high-performance aircraft. The pilot is a Mexican American and the students listen.

Mr. Escalante is determined to teach advanced placement mathematics—calculus—to his class. The other teachers explain that it would be impossible—most of the students are from broken homes and, besides, the school doesn't have the funds for the computers Escalante wants. The teachers and even the students themselves have low expectations.

Mr. Escalante refuses to give up on his dream. He waves off the naysayers, turns aside the threats, finds the funding—overcomes all the obstacles thrown at him.

He insinuates himself into the minds and hearts of his students. He somehow knows what makes them tick. He instills in them the feeling that they are "part of a brave corps on a secret impossible mission," as discussed by Judith Rich Harris in her seminal book, *The Nurture Assumption.*

The students start to work hard on calculus and they learn to support each other. Even the gang leader tries to help out by offering Mr. Escalante "protection" in exchange for two extra textbooks. Escalante accepts the offer, carefully sidestepping direct aggression with the aplomb of a matador.

In the end, eight students take the standardized test and pass with high marks—so high, in fact, that the testing agency suspects the students of cheating due to the statistical improbability of such a feat. The students are forced to take the exam again. Suspicious proctors stare over their shoulders—but the students rise to the occasion and are vindicated. In the ensuing years, Mr. Escalante shares his dream with dozens of other students who stand and deliver in greater and greater numbers.

# 7

## Tell a Story

The importance of storytelling and listening to stories being told can hardly be exaggerated. Stories capture our imaginations, create our myths, and mold our beliefs and values. Stories give our lives meaning; they integrate our brains.

The narrative story teaches children the use and meaning of language itself. It provides vocabulary and a sense of time—a beginning, middle, and end. Story creates purpose.

It has been discovered that some children tell stories to themselves as they lie awake in their cribs, before falling asleep. There is a beautiful story about a little girl named Emily who used more sophisticated language when she talked to herself than when she talked to her parents. Linguists from Harvard, led by Katherine Nelson, studied recordings of young children made on micro cassettes strategically placed in their cribs during a research project called "Narratives from the Crib."

Some of the results are recounted in Malcolm Gladwell's national bestseller, *The Tipping Point.* In her nighttime monologue, two-year-old Emily created a story of her perfect Friday with details of breakfast, kisses from Dad when he went to work, her Nursery School Day (told in hushed tones), and a visit from her friend Carl who rings the doorbell and rushes in. In her story she refers to herself as Emily: "Carl and Emily are going . . . to ride to Nursery School." Emily is organizing her life into a pat structure, she is mastering her routine, and she even tells a joke about the whole scene and says: "Won't that be funny!"

It seems clear that storytelling is a necessary part of thinking and the development of the human brain. Story creates order out of chaos. It establishes patterns that serve as templates for life. Story structures knowledge, making it memorable and whole. Story takes specific ideas, events, and elements and weaves them into a cohesive, holistic narrative.

Story is so powerful that it can give meaning to individuals suffering tragic mental defects as it did for Rebecca, a patient of Dr. Oliver Sacks (as he describes in his extraordinary book *The Man Who Mistook His Wife for a Hat*). Though her IQ was only 60, Rebecca was able to create meaning for herself by acting in a special theater group. The environs of the stage and the unifying force of the story made Rebecca whole, so that Dr. Sacks observed, "One would never guess that she was mentally defective."

In his book, Sacks develops a concept, based on the work of A.R. Luria, which he calls "romantic science." Romantic science explores the concrete (as opposed to the abstract)—it deals with biographies or "novels" of the individual. It is specific, real, alive, and meaningful, as opposed to generic, symbolic, inanimate, and theoretical. It is story: In story we show (the specific), we don't tell (the abstract). Sacks says, "Young children love and demand stories, and can understand complex matters presented as stories, when their powers of comprehending general concepts, paradigms, are almost non-existent."

In *Cultural Literacy*, Hirsch, Kett, and Trefil state that "educated people must know myths, myths, myths." They say that communicating myths is just as important as history: "The tales we tell our children define what kind of people we shall be."

So not only do stories integrate our individual thinking, they also unify our culture.

But what does all of this have to do with rocket science? A few years ago, I read an amazing article by William B. Scott, entitled "Systems Strategy Needed to Build Next Aero Workforce." It appeared in the May 6, 2002 edition of *Aviation Week & Space Technology*. Scott reports on the concerns of aerospace professionals, government and industry leaders, educators, and physicians that "kids exposed to 'light screens'—television, computers, and videos games—for extended periods at an early age do not develop the sensory pathways that enable imagination and creativity." According to Michael Mendizza, a researcher for Touch the Future, "Before the 1950s childhood had a rich, descriptive narrative as its primary environment—story telling and radio. Descriptive words were used and they demanded a child create a corresponding mental image of what those words meant. He painted his own mental picture."

Scott reports that the vocabulary of the average 14-year-old has dropped from 25,000 words to 10,000 in the past 50 years. The optimum development of imagination and creativity occurs in the first five years, when kids are playing, making up stories, and pretending.

The question our government and industry leaders are asking is "who will imagine our future?" Our ability to explore space, to defend ourselves, indeed, even to survive may depend on our ability to listen to and tell stories.

# 8

## Sleep on It

You need your rest—and so does your brain. No one knows why.

But Bertrand Russell, the great mathematician and philosopher, made a personal discovery worth noting. He found that he could rack his brain for months on a problem—and finally solve it. Then he discovered that he could get away with racking his brain for a much shorter initial period—then stop thinking about it—and, after an incubation period, return to find that his subconscious mind had solved the problem in the same total time. After this realization, Russell's work output and creativity took a big leap, and he continued to benefit from his technique into his nineties.

You don't have to be a rocket scientist to take advantage of Bertrand Russell's approach. Your problem doesn't have to be on set theory or epistemology—it could be a homework problem from school, a home-decorating conundrum, an organizational challenge from the office.

The important thing is that, first, it must be a problem that really matters to you. You must have the desire—the *ganas*—that Mr. Escalante demanded from his students. Next, you should learn as much about the problem as you can. You should be intimately familiar with the issues (the term paper topic, the wall-covering choices, and the personnel resources) even though you don't know the answer. It helps to have a number of very specific questions about the problem. Are you confused about how you are ever going to solve it?

Good!

Confusion is often a necessary part of learning and problem solving. If you are never confused, you probably aren't working on problems that are difficult enough for you.

Finally, the last thing you need is time—especially time to sleep on it. That is why it is important to start on hard problems early, so you have sufficient time for your subconscious mind to work on it.

But what if you didn't get started soon enough? Now what should you do? That big homework assignment is due tomorrow afternoon and you figure you can get to bed early and start on it in the morning. Want the lazy man's crash course? Okay, here it is: Work on the problem tonight. Force yourself to carefully read the assignment and struggle hard to really understand what the problem is and what the answer might entail. Then go to bed.

When you wake up, start on your problem again. You should be amazed how much easier it seems, and in many cases you will know the answer or at least what to do next.

# 9

## Think JFK

The most egregious goal ever set is that of President John F. Kennedy:

> I believe that this nation should commit itself to achieving the goal, before this decade is out, of landing a man on the moon and returning him safely to the Earth.

When Kennedy made this statement in May 1961, the United States had very little to go on. We had no micro electronics, no portable computers, no deep-space communications network, no giant rockets, no lunar navigation system, and practically no manned space flight experience. America had yet to put a man in orbit around Earth, let alone go to the moon.

Alan Shepard had flown a suborbital hop that lasted a grand total (from launch to splashdown) of 15 minutes. He spent only 15 seconds in space, near the apex of his 115-mile-high arc. Two months later, Gus Grissom flew a nearly identical suborbital flight with one significant difference. Upon splashdown, his capsule sank to the bottom of the ocean, and Gus nearly drowned.

John Kennedy dreamed a great dream and fired the imaginations of not only Americans but also people around the world. He gave us the big picture and aimed higher than anyone dared believe. Maybe he even BS'd a little. With his New Frontier Program, Kennedy created the desire to reach for space—"to sail this new ocean." He told his nation a story, and Americans embraced it. And when he died, Americans rallied to his dream and fulfilled it.

Does Kennedy deserve credit for getting humans to the moon? He didn't know anything about lunar rendezvous or von Braun's *Nova* rocket. Nevertheless, Michael H. Hart ranked John F. Kennedy number 81 in *The 100: A Ranking of the Most Influential Persons in History*. Hart argued that Kennedy was "the person primarily responsible for instituting the Apollo program." Of course, Kennedy had the help of von Braun, a coterie of the world's most brilliant rocket scientists, brave astronauts, and about 400,000 other scientists, technicians, engineers, and skilled laborers.

And let's not forget the 100 million taxpayers who paid the bill. When Kennedy was speaking about the payload he accidentally said, "It will be the largest payroll—ah payload—in history." Quick to realize his mistake, he added, "And it *will* be the largest payroll." According to Hart, it was Kennedy who made the crucial breakthrough—a political one that required convincing the American public to spend $30 billion to get to the moon.

As long as there are human beings, the achievement of landing people on the moon will be remembered as one of the greatest in history. And it all started with the dream of one man: John F. Kennedy.

# PART II
# Judge

"One cool judgment is worth a thousand hasty counsels. The thing to do is to supply light and not heat."

Woodrow Wilson

# 10

## Get Real

You can dream all you want, but finally you've got to pay the piper—you've got to get real. Look at all that BS you wrote down in your brainstorming sessions—does any of it make any sense? Now you have to separate the wheat from the chaff.

Can we even dignify what you have done before as "thinking?"

Yes it was thinking. There are different types of thinking. In their insightful book, *The Art of Thinking*, Allen Harrison and Robert Bramson describe five distinct types of thinking. They go beyond the simple dichotomy of left brain, right brain in their more specific classification scheme. On the far right they have "the Synthesist," the far left "the Realist," and in the middle "the Pragmatist." Between Pragmatist and Synthesist is "the Idealist"; between Pragmatist and Realist, "the Analyst."

If you are a Synthesist, you probably enjoyed Part I of this book, "Dream." The Synthesist likes to play with ideas, to make things up, to deal with chaos, and to link disparate concepts. If you are a Realist, you might have felt uncomfortable with all the BS. (The hard Realists probably stopped reading by now—but I hope not. Maybe you skipped ahead to the good part.) The Realist, as the name implies, likes to deal with unvarnished reality, the hard facts, concrete ideas. (The other thinking styles will be discussed in more detail later.) Each of the five types represents important thinking strategies; the best rocket scientists use them all.

Most people do not fall into the pure Synthesist or pure Realist categories, but there is a tendency for individuals to favor one method over another. If you are a Realist (and are still with me), then this is the part of the book you will enjoy best. You are ready to cut into that long brainstorming list with all those crazy ideas and set things straight. You don't want theory, you want results! You like ideas that work, and (even if you are a rocket scientist) you consider yourself a down-to-Earth person. Yes, imagination is a lot of fun (perhaps), but you really want to get to Mars.

So go ahead, start tossing the fuzzy ideas off the list and see if there is anything real that you can salvage when you're done. Be careful not to be too aggressive, though. Be thoughtful in your analysis. Be willing to give some new ideas a chance.

And be willing to encourage your Synthesist colleagues or boss to generate more ideas. You might find working with a dreamer to be annoying, even painful. It is typical to find Realists and Synthesists at odds with each other. Read more about each type in Harrison and Bramson's book. Just understanding that there are different types of thinking—and they are all valuable—should help you to get along.

Even more important, learn to recognize and develop the five thinking skills in yourself. Then you will be thinking like the greatest of rocket scientists.

# 11

## Play Games

One way to get real is to create a game out of your problem. (Later we'll talk about the generalization of this idea, which is called simulation.) In the movie *War Games*, a high school computer whiz (played by Matthew Broderick) hacks into a U.S. military war simulator to play "Thermonuclear War." The kid doesn't know it's not a game, and he inadvertently starts World War III. The plot of the movie is based on well-established mathematics called "game theory."

The potential destruction that nuclear war could unleash is so vast that it is difficult to fathom. It is a difficult subject to think about for many reasons. Stanley Kubrick decided that nuclear war was such a depressing subject that he made his classic film, *Dr. Strangelove: or How I Learned to Stop Worrying and Love the Bomb*, into a satire. (I place it no. 6 on my Greatest Sci-Fi Films of the Twentieth Century.) Sometimes it is easier to tackle a profoundly serious subject with a good dollop of humor.

In the 1980s, I thought a lot about the problem of nuclear warfare but made no progress in understanding it until I made a game out of it. (My main interest was to try to understand the mechanics and strategies involved so I could answer for myself what the future is likely to hold for humanity. The intuitive answer—total annihilation—was just not satisfying.)

My personal feelings are roughly expressed by Major Kong, the B-52 pilot in *Dr. Strangelove* who gave a little pep talk to his crew just before delivering his fifty-megaton bomb to Russia: "Heck, I reckon you wouldn't even be human beings—if you didn't have some pretty strong personal feelings about nuclear combat!"

To create my game, I took the popular board game, *Risk*, which is essentially a WWII war game, and bumped it up to include nuclear weapons. (I hope Parker Brothers will forgive me.) By modifying just a few rules, I emulated the effects of nuclear build-up, nuclear shielding, nuclear waste, and post–nuclear war combat.

I was anxious to test my Nuclear *Risk* game with some intelligent players: I found a navigator at the Jet Propulsion Laboratory and a physics professor at Caltech. We played our first game on April 16–17, 1988.

What I discovered after several hours of (simulated) nuclear combat was somewhat obvious in hindsight. My game was more of a test of human psychology than of nuclear strategy.

Here's my summary:

1. As soon as one nation gets one nuke, it is (usually) used on another nation (which, of course, has no nukes).
2. When two nations have nukes and a third does not–then nukes are used (almost) exclusively against the non–nuclear power.
3. If a nation survives a nuclear attack, it launches a counter nuclear attack as soon as possible.
4. In the unusual case when all nations have nukes but nukes have never been used, then WWII style combat continues in a struggle for territory.
5. Whether nukes are used or not, territorial conquest is only achieved by conventional arms. Nukes only destroy, they do not conquer.
6. Once all nations have stockpiled enough nukes to destroy the entire world, the state of MAD (mutually assured destruction) is reached. It is a test of the sanity (or patience) of the players whether to push the button to end the game.

One insight I gained from this experiment is why, during the Cuban missile crisis, some of the U.S. generals advised President Kennedy to launch an all-out nuclear attack against the Soviets before they got an arsenal of nukes. As U.S. Air Force Chief of

Staff General Curtis LeMay put it, "The Russian bear has always been eager to stick his paw in Latin American waters. Now we've got him in a trap, let's take his leg off right up to his testicles. On second thought, let's take off his testicles too." The reasoning of the generals, which sounds insane (and it is indeed horrific), makes sense from a purely game-theory approach. JFK didn't take their advice—and we're still here to talk about it.

Another insight is that if nukes are ever used again, they are likely to be used against a non–nuclear power.

The most important conclusion is the most obvious. It was also "discovered" in *War Games* by the computer: Thermonuclear war is a game that is not worth playing because there can be no winners.

To this we can add—only losers. But you already knew that.

# 12

## Simulate It

In the dream phase, you gave free rein to your creative imagination. Now you must judge your ideas to see if they have real value.

Rocket scientists simulate space missions with computers linked to actual hardware and to mock-ups (models of the spacecraft). To simulate means to imitate the real thing. So rocket scientists do what we see children do all the time: they pretend. When it comes to space exploration, this act of pretending can be very sophisticated and very expensive. Full-scale mock-ups are built. Inside the cockpit are instrument panels and joysticks, outside are projection screens showing views of outer space, the moon, or Mars. When the astronaut moves the joystick, the instrument panel indicates a change in attitude. Sound effects and motion actuators make the experience seem real. Astronauts in actual space flight have often remarked, "That's the best simulation we've ever had!"

In fact, most of the time the experience the astronauts get in the simulator is far worse than the real thing. The reason is that they practice emergency procedures more than the nominal (expected) mission. Mission controllers make up problems for the astronauts to solve; if the astronauts don't react correctly, it could mean certain death for the crew. In the weeks leading up to the launch of Neil Armstrong, Buzz Aldrin, and Michael Collins for the first manned landing on the moon, the astronauts were "getting killed" in the simulator so often that the engineers started having serious doubts about the flight. But this was a good thing. The hard reality of the dangers of a lunar landing was not softened. Any mistake was likely to be fatal. The mission controllers were not sadists—they were able to imagine a lot of problems and were pretty scared themselves.

Eventually, the *Apollo 11* crew learned self-defense. Like karate students, they learned to parry every blow their trainers threw at them. Finally, they were ready for the real thing.

Rocket scientists don't always need billion-dollar simulators and neither do you. When I was working at the Jet Propulsion Laboratory, I discovered that several engineers had sets of Tinker Toys. Tinker Toys come in a cylindrical can. When you dump the can out, you get a bunch of sticks of varying lengths and a lot of little wooden wheels with holes into which the sticks can be jammed. By connecting the wooden hubs with sticks, kids can make cars, windmills, buildings—virtually anything. And rocket scientists can make spacecraft.

When we planned the maneuvers for the *Galileo* spacecraft, which eventually flew to Jupiter, a question would come up about "dual-spin dynamics" and one of the guys would say, "Wait a minute, let me get out my spacecraft." And he'd pull a Tinker Toy model of the *Galileo* off his desk. To a visitor, we must have looked like a couple of overgrown kids playing with their toys. (And actually that is what we were.) We were playing with our Tinker Toys to simulate the complex maneuver modes of a spacecraft headed for Jupiter.

# 13

## Run a Thought Experiment

The cheapest simulation you can do is to run a "thought experiment." Einstein was famous for his thought experiments. When he was only 16 years old, he imagined looking at himself in a handheld mirror. Then he imagined running faster and faster while holding the mirror out in front of him. "What will happen," he wondered, "when I run as fast as the speed of light?" What would he see in the mirror?

Einstein's contemplation of such experiments eventually led him to discover his special theory of relativity.

Another thing Einstein was famous for was his humor. Consider the following thought experiment in which he "explains" how the radio works.

> The wireless telegraph is not difficult to understand. The ordinary telegraph is like a very long cat. You pull the tail in New York and it meows in Los Angeles. The wireless is the same, only without the cat.

This is a thought experiment that doesn't explain anything. It's funnier still when you realize that scientists really don't understand how empty space transmits radio waves, especially because Einstein's special theory of relativity eliminated altogether the putative ether—which was supposed to carry the radio waves.

Thought experiments are conducted only in the mind and so are very inexpensive and safe. To be of any real use, your thought experiments must be accurate representations of reality. Einstein could do meaningful thought experiments because he had done so many laboratory experiments as a boy and because he understood physics so well.

But you don't have to be an Albert Einstein to do a thought experiment. In fact, you perform thought experiments all the time, when you plan a trip to the store or a sightseeing vacation across the country. In these everyday cases, you imagine what you need, how long your trip will take, and how much it will cost. For a long and complicated trip, you may find it necessary to use more tangible simulation tools. Tracing your route on a road map is an example of a simulation (albeit, no longer a thought experiment).

Calculating travel time between rest points is not so different from calculating the flight path of a spacecraft traveling from Earth to another planet. Jotting down your itinerary is a recording of your simulation result.

When you plan a trip, you understand that "the map is not the territory," but you imagine for a while that it is. (This Zen-like saying was coined by Eric Temple Bell, author of *Men of Mathematics*.) If your map is accurately proportioned to the real territory, then your simulation will give you a realistic value for your total trip time and mileage. If that time or distance is too long or the trip is too expensive, you can decide to cancel your trip or modify it to make it work within your budget. This type of thinking is a simulation. Rocket scientists design missions to outer space using much the same kind of thinking.

When I advise my graduate students (who are, in fact, fledgling rocket scientists), I suggest they do a thought experiment in which they imagine how their project will look when it is finished. They may be writing mission design software or developing techniques to solve problems in celestial mechanics, or they may be searching for trajectories to Pluto.

"Imagine what your algorithm will look like to the user," I'll say. "Does it have all the bells and whistles he or she will want? Is what you're doing now going to result in a program that does everything you hope to achieve? If not—change what you're doing."

Realistically imagining how the program or theory or trajectory will look, before you create it, is a great way to judge if your efforts are worthwhile or if you need to change course. That's not just good advice for fledgling rocket scientists—thought experiments are good for everyone.

# 14

## Know Your Limits

When we talk about judging, about getting real, we're usually talking about limits. In rocket science parlance, we talk about dealing with "constraints."

If Johnny can eat one apple in five minutes, how many apples can he eat in sixty minutes? If you do the math you get twelve apples, but Johnny's mother knows better because she's a realist. She knows he's going to start slowing down on the second apple and his little stomach—which is smaller than his eyes—will probably not accommodate the third apple.

Johnny's stomach has its limits. Johnny might be able to imagine eating twelve apples in one hour, and he might bet his lunch money that he's going to do it—but we know better. (Don't try this bet against Paul Newman, however. In *Cool Hand Luke*, he bet his prison inmates that he could eat fifty hard-boiled eggs, and won— which just proves that you shouldn't underestimate limits either.)

Knowing your limits is an important aspect of thinking about a problem. When you plan a long trip, you have to allow for a number of limits including the size of your gas tank, the speed limit, your budget, even your physical limitations.

In rocket science, these limits are extraordinarily important. The amount of onboard propellant determines how long a spacecraft can continue to operate in space. In the multi-ton (mobile home sized) communications satellites that make MTV possible, the value of one year's worth of propellant is over $100 million. That's just for one hundred pounds of propellant. The satellite itself is worth $1 billion—until it runs out of precious station-keeping propellant, when its value goes down to zero. Then another satellite has to be launched into a 22,000-mile-high orbit. Amazingly, several of these satellites are launched every year.

When Armstrong and Aldrin landed on the moon, they had to deal with some very serious limits. They had to land softly enough to avoid damaging their spacecraft, they had to avoid obstacles, and they had to do it quickly—before their propellant ran out—or they'd fall to their deaths on the lunar surface.

Unfortunately, the *Apollo 11* lunar module missed its landing site by four miles (due to a one inch per second velocity error at the beginning of their descent orbit), and Neil and Buzz found themselves hovering over a boulder-strewn field. They had a fuel gauge that was as inaccurate as the one in your car. (We can go to the moon, but why can't we make an accurate gas gauge?) Mission Control back in Houston anxiously called out, "sixty seconds," their estimate of how much longer the fuel would last. Then, "Thirty seconds." Then they waited and listened. Buzz Aldrin read off altitudes and descent speeds, "forty feet, two and a half down. Picking up some dust."

By this time the fuel gauge was on empty. But you know when your car's gas gauge is on empty how it might still have a gallon left—or it might be bone dry? It depends on the car. The lunar module fuel gauge had an error of about 2 percent—and it was registering empty. They could run out of fuel at any second.

Finally Neil Armstrong called out, "Houston, Tranquility Base here. The *Eagle* has landed."

To which Mission Control replied, "Roger, Tranquility, we copy you on the ground. You've got a bunch of guys about to turn blue. We're breathing again. Thanks a lot."

So the moral of our story is that when you try to solve problems that have all sorts of limits, you are thinking like the rocket scientists and the astronauts who made the first landing on the moon happen. It is easy and fun to say, "Think outside of the box," and there is a time for that kind of thinking. But when you want to get real, you have to stay inside the constraint box—that's where the challenge is.

# 15

## Weigh Ideas

All ideas are not created equal. Some ideas are better than others. When you see a good idea, you recognize its quality immediately. (For a book-long, stirring essay on quality, read *Zen and the Art of Motorcycle Maintenance* by Robert M. Pirsig.)

Lack of quality is easy to spot. Maybe you noticed how poorly designed your coffeemaker is: how the pot dribbles down its side and onto your sock, how hot scalding vapors rise up and burn your hand while you're holding the flimsy plastic handle, how god-awful it is to clean.

This coffeemaker is an example of theory with no practice, a manufactured idea with no discrimination. The coffeemaker might look nice—but it doesn't work. The inventor (if he deserves the name) never used his coffeemaker and it shows.

The "proof is in the pudding" means you have to taste it. Once you put it in your mouth, judgment soon follows.

The Sage of Baltimore, H.L. Mencken, made it a constant theme of his writing that the "weighing of ideas" is the essence of real thinking. The longest reigning chess champion in history, Dr. Emanuel Lasker, once said, "When you find a good move, look for a better one!"

The weighing of ideas, the selection of better over good, is the balancing process that must follow brainstorming. The products of the unfettered imagination are judged coolly and soberly, without prejudice.

When people talk about imagination, they often have very foggy notions of what it is. It is commonly assumed that being imaginative means being fanciful, undisciplined, even flighty. But when rocket scientists use the word "imagine," they mean something more precise. They mean to picture a space mission that is really possible. They don't consider using antigravity, or warp drive, or pixie dust as a means to deliver their next spacecraft to Saturn. They have what I call an "accurate imagination." They temper their fondest dreams with a cold splash of reality. That's because rocket scientists want to live in the real world, not the virtual world. Sure they love science fiction and video games, but more than anything, they want the real thing.

Rocket scientists know, as Carl Sagan said, that "space is a place." They really want to go there and they know they can. But they're not depending on the *Star Trek* transporter to beam them there. John Kennedy would understand these rocket scientists. He said of himself, "I'm an optimist without illusions."

# PART III

# Ask

"The 'silly' question is the first intimation of some totally new development."

Alfred North
Whitehead

# 16

## Ask Dumb Questions

"The only dumb question—is the one that isn't asked." That's what I was told when I started working at NASA's Jet Propulsion Laboratory (JPL). I quickly learned that all these brilliant scientists and engineers had a culture of questions: there's no such thing as a dumb question.

During the planning of the Mars Climate Orbiter, the dumb question that wasn't asked was, "Are these numbers in metric or in the English system?" The unasked question of centimeters versus inches led to the destruction of the $200 million spacecraft when it dove into the Martian atmosphere and exploded.

Remarkably, at Caltech, which operates JPL, the students have adopted a custom that is the antithesis of the JPL culture. The undergraduate class consists of the most intelligent students in the country with an average IQ of 150. The professors (among them several Nobel laureates) who teach there rarely hear a question from the students. Nobody wants to ask anything. Why not?

I observed a similar phenomenon when I was working (long distance) with Dr. Buzz Aldrin (the lunar module pilot for the first human landing on the moon). I arranged for my students to make teleconference presentations to Dr. Aldrin about the research we were doing for him on a human transportation system that would cycle between Earth and Mars. After our second meeting, I realized that my students never asked any questions of Dr. Aldrin. I asked them why. "We were afraid to appear stupid," said my top doctoral student.

The answer was fear.

There is a tendency for older adults to care less about what others think of them, but students can be strongly influenced by (often nonverbal) peer pressure. (If you have any doubt of this, consider Judith Rich Harris's insightful book *The Nurture Assumption: Why Children Turn Out The Way They Do*.) The fear students have is usually based on how their classmates might react. I notice that when students privately ask me questions (after class or in my office), this fear disappears. They don't care what I think of them.

When he worked on the atomic bomb in Los Alamos, Richard Feynman was asked by a general to review the safety of the new designs for the Oak Ridge plant. The plant was to separate isotopes of uranium—the nuclear fuel for the bomb. Two engineers rolled out a complicated blueprint with many symbols that Feynman could not decipher. The engineers had boasted that they had redundant valves everywhere so that if any one of them failed, a secondary valve would prevent an accumulation of uranium—a potentially explosive situation.

Completely flummoxed and unsure whether the "X" he was looking at was a valve or a window, Feynman stabbed his finger at the blueprint and asked, "What if this valve fails?"

The engineers looked at each other and thought for a moment. Worried looks appeared on their faces and one of them said, "You're absolutely right, sir!" Then they excused themselves to examine the problem further—dire consequences were indicated.

Then the general, who had invited Feynman to study the plant design said, "I knew you were a genius when you spotted that valve problem!"

# 17

## Ask Big Questions

Carl Sagan wasn't ashamed to ask big questions. "How did life begin on Earth?" "Can we duplicate those conditions in the laboratory?" "Does life exist on other worlds?" "Is there intelligent life in the universe?" "If so, how can we communicate with them?"

Carl Sagan asked a lot of big questions for a scientist. Most scientists concentrate on small, highly specialized questions and use reductionist techniques to make progress. But Sagan was a generalist. He knew as much about biochemistry as he did astronomy— and he yearned to be a rocket scientist too. As a boy he loved science fiction. As a scientist he was prolific. (For two fascinating accounts, see *Carl Sagan, A Life* by Keay Davidson and *Carl Sagan: A Life in the Cosmos* by William Poundstone.)

Sagan also wrote the popular science fiction novel *Contact*. In 1980, Sagan became the showman of science when he broadcast his *Cosmos* series. I was amazed to see the reactions of my friends and colleagues at JPL to his show. The secretaries, administrative staff, and the technicians loved Sagan's program. "Now I understand what we're doing here!" one of the staff exclaimed with glee.

But the scientists and engineers took a dim view of *Cosmos*. "Bad acting, bad writing, bad science!" they'd say. I found it perplexing. I couldn't understand why they were so angry with Sagan. What was wrong with popularizing science? Were they jealous that he had become so famous for appearing in a TV series?

There was a clear dichotomy. Nontechnical people thought Sagan was great: scientists thought he was terrible. He wasn't quite an abomination—just an embarrassment.

Sagan paid for his generalist tendencies. He was denied tenure at Harvard; his nomination to the National Academy of Sciences was rejected. Scientists and engineers laughed at him behind his back. Although he published more than four hundred scientific and popular articles—more than the average Nobel prize winner—he was not considered a true scientist by the academy.

Asking big questions is the mark of a Synthesist—one who puts disparate concepts like biology and astronomy together. Most scientists are Analysts—who believe there is one, right way to do science. Asking big questions—out loud—is something akin to asking an astronaut about the "Right Stuff." It just isn't something you talk about. (All of this is explained by Tom Wolfe in his extraordinary book *The Right Stuff*, which was also made into a great movie.)

This theme of "who are you to ask?" is nicely played out in the "Galaxy Being," the first episode of the original television series *The Outer Limits*. A scientist, Alan (brilliantly portrayed by Cliff Robertson), is draining power from his commercial radio station to explore space. His wife, Carol, wants to know why. And how they are going to explain to their sponsors that their radio commercials have disappeared into a feeble beep. After an enigmatic pause, Alan says, "Because it's interesting." Carol becomes frustrated and asks, "What makes you think that you can discover anything? Who are you?"

"Nobody," answers Alan. "But the secrets of the universe don't mind—they reveal themselves to nobodies who care."

"But now they have big laboratories that work on all those things," says Carol.

"The big laboratories spend millions of dollars, Carol, and they work slowly and surely and may get results, but not the big steps, not the breakthroughs—they come from the human mind—not the laboratory."

Carl Sagan would probably have enjoyed that *Outer Limits* episode. Regrettably, Sagan died prematurely of pneumonia (after a two-year battle with a preleukemic bone marrow disease) in 1996. His contributions to science are yet to be fully appreciated. Sagan gave dignity to those big questions he liked to ask. New generations of scientists are not afraid or embarrassed to ask them anymore. Sagan's legacy is built into the current NASA mandate for exploration, which now includes: to understand the origins of life, to search for life, and to seek evidence of intelligent life in the universe.

Thanks, Carl!

# 18

## Ask "What If?"

"They're all a bunch of what-iffers over there at the lab," a Caltech professor's wife remarked. "They might as well ask, 'What if the sky should fall?' as far as I'm concerned."

Her disdain for "What-iffers" is a common reaction. People who ask a lot of questions can be quite annoying—particularly if the questions are good ones. The laboratory the lady was referring to was Caltech's Jet Propulsion Lab—NASA's crown jewel. The "bunch" were all the rocket scientists working there.

It's true—there is a worrywart side to rocket scientists. And for good reason: Murphy's law. "If anything can go wrong—it will." Learned from hard experience. Space travel, after all, is extremely hazardous. It involves riding a highly explosive rocket (essentially a flying bomb) into orbit, living in the space environment with its dangers of airlessness, microgravity, and radiation, and then surviving a fiery reentry through Earth's atmosphere. A lot can go wrong—and a lot has. You'd be a fool not to ask a lot of questions. This is not fear—it is reason.

Here are some what-ifs we dealt with along the way to the moon.

Question: What if the rocket blows?
Answer: Use an escape system that catapults the astronauts high above the explosion and deploys parachutes to save them.
Question: What if the rocket comes crashing down in a residential or tourist area near the sunny beaches of Florida?
Answer: Detonate the self-destruct system—blow the rocket to kingdom come.

Question: What if the Russians should get there first?

Answer: Build the *Apollo* spacecraft—but fast—and win the race. Our very survival is at stake and failure is not an option here.

Question: What if Alan Shepard has to pee?

Answer: Didn't see that one coming—let him pee in his suit.

Question: What if Gus Grissom has to pee?

Answer: Got that one covered—put a rubber on him.

Question: What if John Glenn's heat shield should detach before or during reentry?

Answer: Don't jettison the retro-rockets—the straps may hold the shield on just long enough. (The right answer, but it turned out to be a false alarm—a faulty warning light. But it's good for us, keeps everyone on his or her toes—especially Glenn.)

Question: What if the *Apollo* loses power on the way to the moon?

Answer: Do we have to think of everything? Oh—that's right—we do. Use the lunar lander as a lifeboat. Good thing we thought of this—it saved the *Apollo 13* crew and made a pretty good movie. (See the film *Apollo 13*—it's great!)

We had a nice collection of what-ifs during the *Apollo* days. Unfortunately, over the next three decades NASA (in a sort of institutional Alzheimer disease) forgot the hard-won lessons of its youth. Consider some of these unanswered, or poorly answered, what-ifs that apply to the shuttle program.

Question: What if the shuttle blows during launch?

Answer: First of all that's an unfair question, because it's not going to happen. We estimate that the chance that a shuttle will be destroyed during a launch is 1 in 100,000 launches.

Question: But, really, what if it does blow?

Answer: Then the astronauts die.

Question: What if the shuttle damages or loses its heat shielding before or during reentry?

Answer: Again we're talking about an extremely rare event. We estimate the odds of that happening to be infinitesimal—zero actually.

Question: But what if it does, say, lose a bunch of tiles?

Answer: Then the astronauts die.

Question: What if the shuttle guidance fails during reentry—say they lose power?

Answer: Hasn't happened.

Question: But if?

Answer: Then the astronauts die.

Question: What if we keep flying the shuttle, knowing that it has so many failure modes?

Answer: Well, space travel is not for the fainthearted. You've got to expect a few accidents, maybe a few fatalities. But if we didn't accept this, then we couldn't have a shuttle program.

Question: What if we can't get to the space station because the shuttle is too unsafe to fly?

Answer: Well, this line of questioning has been highly speculative, but if we accept this hypothetical case—then we're talking about going to the Russians and paying them to get us to the station.

Question: What if the space station didn't exist?

Answer: Then the shuttle would have nowhere to go.

I hope I don't sound too negative. Let's add a positive note to finish off:

Question: What if we decided not to complete the construction of the space station, but put those funds toward real space exploration?

Answer: Then we would save about $100 billion, which is more than enough to send the first humans to Mars!

# 19

## Ask: "Animal, Vegetable, or Mineral?"

There's an old game called "Twenty Questions" in which one person thinks of the name of an object and the other person, or group, tries to guess what it is. The first question is, "Is it animal, vegetable, or mineral?"

After that, all questions must be posed as "yes" or "no" questions.

The answerer must be knowledgeable about the object she has selected and scrupulously honest in her answers. Sometimes she may be uncertain and should say so or ask "Can you reformulate your question?"

A version of this game, called "What's My Line?" was a favorite TV program during the 1950s and 1960s. The host would present a new guest each week, and the panel of five regulars—who were masters of the game—would ask yes-or-no questions to determine the guest's line of business. Occasionally, the guest would be puzzled and there would be a whispered conversation with the host to clear things up. There was never any intention to mislead the players.

A whole chapter of Rudolf Flesch's excellent book *The Art of Clear Thinking*, is devoted to "Animal, Vegetable, or Mineral." Incidentally, Flesch is also the author of the famous book *Why Johnny Can't Read*.

Flesch states that "Twenty Questions" is *the* model of productive thinking.

I read *The Art of Clear Thinking* in 1974, as a first-year graduate student and started practicing "Twenty Questions" with one of my classmates, a brilliant student in aerospace engineering. The idea of the game is to ask questions that will divide the "universe" of possible answers in half. So, for example, after being told that the object is mineral, a good follow-up question might be: "Is it man-made?" This question divides the universe of answers into all those objects that are natural in origin and all those that are created by human beings. These are not exactly equal in number, but we can say that, roughly, we have divided the number of potential objects in half. If the answer is yes, we can stop thinking about quasars and quarks, protons and planets: We know the object is of human origin.

Once my friend knew it was a man-made object, he had a great follow-up question: "Can you buy it at Thrifty Supermarket?" I suppose today we would say "Wal-Mart." These are supermarkets that have everything from automobile parts to Zoloft. But there are still many things made by people that aren't in the store. Consider that vast number of products produced by the military–industrial complex. You can't buy a hand grenade at Wal-Mart. So the Thrifty question was a good divider.

Even neater was the series of questions that would follow if I said yes to Thrifty. Then my friend would divide up the store into departments, as if he were taking a walk through the aisles. This is easily accomplished by grouping departments: "Is it in any of the following categories—automotive, hardware, housewares, or electronics?" Notice that this is a fair yes-or-no question. You don't answer "Yes—housewares, boy are you clever!" You simply say yes, without telling the questioner which of the four is right. He will have to ask two more questions at most to differentiate between the four objects.

"Is it housewares or hardware?"

"Yes."

"Hardware?"

"No."

Then the questioner knows it is housewares and doesn't waste another question confirming it.

The perfect questioner will divide the potential objects in half up to twenty times in a row, which means that she can differentiate $\frac{1}{2} \times \frac{1}{2} \times \frac{1}{2} \ldots \times \frac{1}{2} =$ one out of about a million objects. Because the English language has about a million words (many of which are not objects), the perfect questioner should win all games in which the object is described by a single word.

Of course, a very poor question to ask, after determining that the object is man-made, would be, "Is it a thumbtack?" If there were one million objects, then when you are told "No," you have not narrowed the question much: there are now 999,999 objects left to choose from.

The challenge of the game is that you have to think in terms of categories, classifications, and hierarchies. You must be imaginative and use good judgment. You must have a grasp of the meaning of words. "Twenty Questions" is a wonderful game that improves your thinking and costs nothing to play. It can be used to generate ideas and sharpen your understanding of things.

To make your games more enjoyable, I recommend writing down each question and answer and having a referee to keep the answers clear and honest. If you can play "Twenty Questions" well, then you're thinking like a rocket scientist.

# 20

## Ask Just One More Question

One of my favorite (non–sci-fi) TV programs is *Columbo*. Columbo (played endearingly by Peter Falk for three decades) is a short, disheveled detective in a crumpled raincoat who solves homicides for the LAPD (Los Angeles Police Department).

Columbo doesn't look like a rocket scientist.

Each episode is a detective who-dunit story, only we know who because we are shown the murderer committing the crime in the first scene. So what's the point of the story?

The murderer is typically one of the rich and famous, an arrogant, powerful man or woman who thinks he or she can get away with it. After all, these people are a lot smarter than the LAPD. When the murderer first encounters the frumpy and disorganized Columbo, the killer is even more convinced that he or she has nothing to worry about.

The idea of each episode is: How can Columbo prove (to anyone but the most obtuse jury) that the murderer did it? One thing we know about Columbo is that his instincts are very keen—he starts asking the murderer a lot of pointy questions. He knows who did it—but how?

The fun in watching Columbo is to see how he struggles with a series of questions that pile up in his brain and foment confusion. Why is he confused? Because he is a very logical man. He requires consistency in his universe. Inconsistencies bother him. Little things that no one else would notice disturb him. "How could the gun fall on top of a dried blood drop, if this were a suicide?" "Why is there liquid water in the freezer compartment of the refrigerator?"

Columbo believes that the universe and human beings behave according to a consistent set of rules. If there is an apparent violation of that consistency, then he wants to know why.

Columbo would play out the crime in his mind—a kind of Einstein thought experiment—then he'd get stuck. His mental simulation would break down. Something didn't make sense. But in Columbo's world, everything had to make sense.

He would share his confusion with the killer. Remarkably, the killer would provide an ingenious solution to Columbo's dilemma. "During the suicide the gun hung up on the dead man's hand—for a moment. A kind of rigor mortis convulsion would spring the index finger open—allowing the gun to drop—after the blood had dried."

*Voilà!*

And Columbo would be grateful for the explanation. "Thank you, sir! That clears it up. You know I was really going in circles on this one, but you really straightened me out. Thank you very much."

Then Columbo would go away and the murderer would breathe a sigh of relief. (So would the audience. By this time we were secretly hoping the criminal would get away with it.)

A moment later, Columbo would burst through the door with, "Just one more question, sir. I really hope you don't mind. I almost forgot."

Of course the killer is someone who is very important—a politician, a famous conductor, a wealthy business woman—on a very busy schedule. He or she doesn't have time for all this nonsense.

In every episode the murderer assumes that Columbo is a fool. He looks like a fool in his wrinkled raincoat and he sure asks a lot of questions. Why does everyone assume that a person who asks a lot of questions is stupid?

In the end, we learn (again) that Columbo is a genius. The killer runs out of exculpatory explanations—he or she is finally trapped. There is no escaping the logic of Columbo's hypothesis: either you are the murderer or the laws of the universe have been broken. In the final scene, the murderer gives Columbo a wry smile and admits, "I have underestimated you."

"Yes, sir, I think you may have," says Columbo, his eye twinkling.

Columbo thinks like a rocket scientist: he's not afraid to ask "just one more question."

# PART IV
# Check

"The disaster had been looming ahead for many months, and I had studied my plans for all contingencies a hundred times."

Ernest Shackleton

# 21

## Prove Yourself Wrong

"If anything can go wrong—it will." This simple statement of Murphy's law looms large in the rocket scientist's consciousness. In the early days of the American space program, there was another saying, "Ours always blow," which meant that you could count on our rockets to explode every time.

This school of hard knocks (and violent explosions) was particularly difficult for Americans to bear because of the dramatic successes being enjoyed by the Soviet space program. Eventually, our rocket scientists learned that nearly every system had to have a backup system, that all calculations had to be checked and double checked, and that nothing could be taken for granted—except human error.

Rocket scientists are thought to be geniuses who never make mistakes, when in reality they are human beings who make spectacular mistakes. Their errors are cataclysmic, expensive, even deadly. And, at least in the American space program, these disasters were public.

Rocket scientists are intimately familiar with failure. For this reason, they have learned to deal with mistakes—to avoid them if at all possible. They have been humbled by experience and in their new-found humility have learned that they must rid themselves of error—they must find the fault within themselves. They must seek to prove themselves wrong.

This fault-finding is nothing new. It is a basic tenet of science to propose theories that can be tested, that can be proved wrong. Carl Sagan was a strong proponent of the philosophy of Karl Popper: that no theory is scientific unless it can, in principle, be proved wrong. Scientific theories must be testable. Richard Feynman described this attitude of science as one of bending over backwards to prove itself wrong. Only after all attempts to disprove a theory have *failed* do we start to consider the theory a credible one.

So, if you want to think like a rocket scientist, swallow your pride and try to prove yourself wrong.

# 22

## Inspect for Defects

Quality control is looking over that book you're about to buy—for quality of the writing and for defects in its physical production. Every seemingly identical object is not identical. If it's made by human beings there are defects. Maybe you are looking at this book in a store and have decided to buy a copy. Or maybe you have already paid for it, in which case the following exercise may disappoint you.

Take a look at the physical characteristics of the book you are holding in your hand. Check for dog-eared pages, tears in the spine or cover. Flip through the pages and look for stray marks, glued-together pages, and *latte* damage. Close the book and look at it at an oblique angle so you can see the shine off the cover—this will reveal smudges, fingerprints, and dust. Yech! Look all around the edges of the book and notice dents and gouges indicating it has been dropped.

If you find a flaw, then I hope you are still in the store. Pick up another copy (preferably in the back of the display or on the bottom of the stack) and inspect it. Look at several books. If you look closely, you will find that they all have flaws. Pick out the one with the fewest defects.

You have just performed a process that, in the aerospace industry, is called quality *control*. Quality control is kicking the tires, taking a test drive, and looking under the hood (even if you're not sure what you're looking for). You can save yourself a lot of money and grief by simply taking a look.

As Yogi Berra said, "You can see a lot—when you observe!"

Senator Lloyd Benson put it this way: "You can expect what you inspect."

Why should you pay for someone else's mistake? You can use this technique of quality control to improve the quality of your life.

In Mission Control, there are experts in charge of all the major systems. You get to see them in the two great films of space exploration, *The Right Stuff* and *Apollo 13*. Before launch, the flight director calls on his experts—the capsule communicator, the doctor, the flight dynamics officer, the retro-fire engineer, and the range safety officer. These are the quality control experts who constantly monitor the state of the rocket, spacecraft, and astronauts during countdown.

They observe.

If anything is not up to par, the officer will call out, "No go!" when the flight director does his roll call. If "Flight" (the flight director) gets a single "No go!" he will put a "hold" on the launch until the problem is solved. (For the inside story, see the *New York Times* bestseller *Flight: My Life in Mission Control* by Chris Kraft.)

The experts have call names for space-age efficiency, so what you hear just before launch is the flight director calling and the officers responding:

Capcom?

Go!

Surgeon?

Go!

FIDO?

Go!

Retro?

Go!

RSO?

Go!

And we are "go" for launch.

That is one of the ways rocket scientists beat back errors in their attempt to ensure success.

You can use the approach to improve life here on planet Earth. When you find a product or service lacking, let the manager know.

She'll send the defective product back to the manufacturer or chide the lackluster employee. You're not the only one who benefits when you apply quality control (although you may be the first). Companies respond to their customers' expectations—or they go out of business. When you expect the best for yourself, you improve quality for everyone.

Quality control isn't just for rocket scientists.

# 23

## Have a Backup Plan

The quickest way to separate the rocket scientists from the non-rocket scientists is to check out their backup plan. All you have to do is ask, "So what's Plan B?"

Space travel is so difficult and dangerous that there have to be doubly and triply redundant systems to beat back catastrophic errors and to maintain a level of safety. Rockets have a nasty habit of exploding—after all, 90 percent of their weight is composed of highly combustible propellant. They are for all practical purposes flying bombs. (But that's what makes rocket science so interesting.)

In the beginning, our flying bombs did what you'd expect them to do—they exploded half the time. Then, after some experience, our rocket scientists got them to explode only 10 percent of the time, which was cause for great celebration and jubilant merrymaking. Today, after a half century, even our best rockets still blow up about 2 percent of the time. That's one out of every fifty flights.

The average launch vehicle costs about $500 million and the average communication satellite is worth nearly a billion dollars. If it were possible to keep our rockets from exploding 2 percent of the time, we would have figured that out a long time ago because the economic incentive is tremendous. (We might even say "astronomical.")

Rocket scientists realized in the beginning that for human missions, we would have to employ escape systems to save the lives of the astronauts in the event of a launch mishap. An escape tower was attached to the top of the *Mercury* capsule that would pluck the capsule off the top of the launch vehicle, carry the astronaut thousands of feet above the exploding rocket, and then deploy parachutes to ensure a safe landing. A similar system was used on the three-man *Apollo* capsule; an ejection system was used on the two-man *Gemini*.

Unfortunately, many of these lessons were forgotten or ignored in the shuttle program, which is why seven astronauts were killed in 1986 in the *Challenger* launch. The booster exploded seventy-three seconds after launch and the astronauts did not eject nor did they use an escape tower—because they didn't have ejection seats or an escape tower. These backup systems were deemed too heavy to include in the shuttle, which was supposed to, according to NASA, have a 1 in 100,000 chance of failing during launch.

Nobel laureate Richard Feynman compared shuttle flights to "playing Russian roulette . . . you pull the trigger and the gun doesn't go off, so it must be safe to pull the trigger again. . . ." (For a detailed account, see Feynman's brilliant book *What Do You Care What Other People Think?* In Part 2, "Mr. Feynman Goes to Washington: Investigating the Space Shuttle *Challenger* Disaster," he tells the story as no one else can.)

The bottom line is that after performing the greatest technological feat in the history of the human race—landing men on the moon—NASA stopped thinking like rocket scientists and built the shuttle.

# 24

## Do a Sanity Test

"Doing a sanity test" is rocket science parlance for simply asking, "Does this make sense?"

Rocket scientists can become myopic in their work. (After all, most of them are near-sighted and have to wear thick glasses.) They get caught up in their equations, calculations, and computer simulations. They can become so involved in the mathematical details and technical minutiae that they lose sight of the big picture—like an instrument-rated pilot glued to his gauges who never looks out the window to see where he's going or if he's about to collide with another aircraft.

When a rocket scientist does a sanity test, he's pinching himself back into reality and asking if what he has done adds up. One of the challenges that rocket scientists experience is that they live in a world of mathematical symbols that they manipulate in order to understand how the rocket will perform in space. These symbols are shorthand for the laws of the universe. By writing equations and sorting the symbols by the rules of algebra, the rocket scientist can predict how high and how fast the rocket will go and when it will get to its destination, which could be as far away as Pluto—4 billion miles from the sun. For example, when the *Voyager* spacecraft arrived at Neptune (which is nearly as distant as Pluto), the error in the spacecraft's arrival time was only fourteen seconds out of a total trip time of nearly twelve years.

So the laws of celestial mechanics are extremely precise.

But there is a downside to this world of symbols: The slightest error in addition or algebra makes the entire analysis wrong—completely meaningless and without value. The same thing applies to the rocket scientist's dependence on the computer: one wrong keystroke and everything is wrong. Garbage in, garbage out!

Thus, rocket scientists have to constantly look up from their desks or computer screens and ask, "Does this make any sense?"

The mathematics that rocket scientists wield is the key that grants them access to the secrets of the universe. Mishandled, it can produce delusions and disaster.

Indeed, sanity tests need to be performed regularly and not just by rocket scientists.

# 25

## Check Your Arithmetic

"Measure twice; cut once," is the carpenter's adage.

Now carpenters have been around for several millennia, so we can say that rocket scientists are thinking like carpenters when they check their calculations.

When Albert Einstein finally published his general theory of relativity in 1915, he was at his wit's end because the world's greatest mathematician, David Hilbert, was in a race with him to find the correct theory. Einstein had spent ten years "going down blind alleys" trying to formulate his theory of gravitation. Each year he published a new version of the theory—and the next year he would recant and propose a new one. He struggled with the monstrous mathematical machinery of tensor calculus (mathematics so recondite that even Isaac Asimov—the "Great Explainer" and author of nearly five hundred books—admitted he could not master it).

Einstein wrote hundreds of pages of calculations in tensor calculus. He was twenty-six years old when he started his quest. When he finished the theory at the age of thirty-six, his health was shot and his hair was gray. The effort nearly killed him.

But if he had been more careful, Einstein would have saved himself a lot of pain. When he reviewed the calculations he had done in 1913, he found he had made an error, which when corrected gave him the final theory! He had had the right solution in his notes for two years and didn't realize it.

So some of the greatest geniuses make costly mistakes. Keep this lesson in mind when you have to balance your checkbook or calculate your income tax. You are not the only one who finds such calculations tedious and frustrating. Everyone makes mistakes.

But the checkbook must be balanced, the income tax must be paid, and errors will cost you. If you make a mistake in your checkbook, you could end up overdrawing your account, getting fined by your bank, and losing your credit. A mistake with the Internal Revenue Service could cost you significant penalties, including the possibility of criminal charges.

Dire consequences undoubtedly add to the fear and frustration of checking your arithmetic. Perhaps it will help if we think of Albert Einstein's plight. The great man who said, "The most incomprehensible thing about the universe is—that it *is* comprehensible," found the instructions he received from the IRS impossible to follow!

# 26

## Know the Risks

When people say, "They knew the risks," they usually mean, "They knew they could be killed."

But knowing you might die is not the same thing as knowing the risks—it is only knowing a possible outcome. What "knowing the risks" really means (to rocket scientists) is knowing the numbers, the probability that you might die. For example, the probability that you (or I) will be killed in an automobile accident is one in eighty, a little over 1 percent. This is the average number over the course of a lifetime.

Each time you board an airliner, you take a risk of about one in a million of dying (about the same as for each car trip). Astronauts have a one in fifty chance of dying during a shuttle flight. Knowing the risk—the probability—is the first step to dealing with it.

Now you may well ask: "How can we know a number like that?" A fair question. In many cases, the number is very well known because of the large amount of data. The risk of death in an automobile accident is well established by insurance actuaries who calculate insurance premiums. They must accurately assess the risk in order to offer competitive rates, while ensuring a profit. And they do it very well. (Otherwise, the insurance companies could lose their shirts.)

It is often said that you can lie with statistics. But—it's even easier to lie without them. When statistics are used correctly, they can bring us closer to the truth.

In the case of space exploration, the probabilities are more uncertain because of the paucity of data. The statistics of 100 million drivers give insurance actuaries far greater confidence than the statistics of hundreds of space missions give rocket scientists. But rocket scientists have another trick up their sleeves: failure analysis. Failure analysis is a branch of mathematics that can be applied (in a study of the myriad components of a single rocket) to determine how often the rocket will blow up. A great deal of space mission planning depends on such probabilistic models.

The U.S. Air Force did a failure analysis of the shuttle, well before the *Challenger* crashed in 1986. They estimated a launch failure rate of 1.5 percent at the time that NASA touted a failure rate of 1 in 100,000. Upon hearing the NASA number, Richard Feynman commented: "That means you could fly every day for 300 years without seeing a crash." It defied common sense.

Aerospace engineers have understood and applied risk assessment for many decades. The reason that airline travel is so safe is because they really know the risk and have done something about it. For example, the landing gear in an airliner has a failure rate of about one in a thousand. However, all airliners carry at least two independent backup systems that each have the same failure rate. By the rules of probability, the chance that all three systems would fail is one in a billion. That is precisely why you rarely hear of an airliner crashing due to landing gear failure.

Rocket scientists understood the risks—knew the numbers— and made sure that the *Mercury*, *Gemini*, and *Apollo* astronauts had viable escape systems to save their lives in case of a launch failure. And as we have discussed earlier (but it bears repeating), these lessons were forgotten or ignored when the shuttle was built.

In interplanetary space exploration, rocket scientists knew that in order to ensure the success of their robotic missions, they should build twin spacecraft. They realized that it was not as expensive to build a second, duplicate spacecraft because most of the cost was in the design of the first spacecraft, and they understood that the chances of success were much better with two spacecraft instead of one.

This commonsense approach was applied in the *Mariner* missions to Venus and Mars and really paid off. *Mariner 1* and *Mariner*

*2* were twin spacecraft designed to explore Venus. On December 14, 1962, *Mariner 2* succeeded in confirming that the cloud-enshrouded planet had a surface temperature exceeding 800 degrees Fahrenheit, as predicted by Carl Sagan. *Mariner 1*, launched previously, disappeared into the Atlantic Ocean.

*Mariners 3* and *4* were sent to explore Mars. *Mariner 3* crashed due to a launch failure, but *Mariner 4*, launched in November 1964, succeeded in sending back twenty-one (and a half) pictures proving for the first time that Mars had craters like those on the moon.

*Voyagers 1* and *2* were launched in 1977 to explore Jupiter and Saturn and to take advantage of a planetary alignment of two other planets, Uranus and Neptune, a rare event occurring every 175 years. This planetary alignment gave rise to the term "Grand Tour," meaning that four planets could be reconnoitered by a single spacecraft.

The nominal mission was to explore Jupiter and Saturn. If *Voyager 1* failed, then *Voyager 2* would serve as backup, giving up the Grand Tour in the process. On the other hand, if *Voyager 1* succeeded, then *Voyager 2* would be targeted toward Uranus and Neptune, taking advantage of a gravitational slingshot off of Saturn. In the case of the *Voyager* missions, both spacecraft were spectacularly successful, and the scientific return was far greater than that of the nominal Jupiter–Saturn mission.

Because rocket scientists were conservative about risk, they overbuilt their spacecraft—and those spacecraft sometimes exceeded all expectations. Another example was the *Viking* mission, which searched for life on Mars. In July 1976, *Viking 1* landed on Mars and operated flawlessly. *Viking 2* followed suit in August. The only failure was the failure to detect life. (But it's not the spacecraft's fault if there isn't any life there.) Both spacecraft performed all the biological experiments. They survived on Mars for years, far beyond their mission plans, and were eventually turned off due to lack of funding to continue monitoring them.

In the 1980s, these lessons of redundant robotic spacecraft were promptly forgotten and NASA built a single, extraordinarily complex spacecraft, the *Galileo*, to orbit Jupiter. JPL mission planners argued the benefits of building a twin, launching the first spacecraft to Jupiter, and if that mission was successful, launching the second craft to Saturn. NASA headquarters rejected this suggestion (because of the expense), and only a single spacecraft was built to orbit Jupiter. The *Galileo* spacecraft very nearly failed when its high-gain antenna (used to transmit high-definition pictures and to navigate the craft) failed to open. Thanks to the creative engineering pulled off by the mission designers at JPL, the mission was a success, albeit with far fewer images of Jupiter and its moons.

Recently, NASA has returned to the practice of building twin spacecraft. After two 1999 missions failed (the Climate Orbiter and the Polar Lander), NASA sent twin robotic rovers, *Spirit* and *Opportunity*, to Mars. Both rovers landed in January 2004 and were highly successful. About the same time, a Japanese spacecraft, the *Nozomi*, and a European probe, *Beagle 2*, both failed in their missions to Mars. These failures are a reminder of the high risk in space exploration and of the importance of employing redundant spacecraft to combat the risk.

# 27

## Question Your Assumptions

In the 1976 movie *The Bad News Bears*, the coach (Walter Matthau) points out the folly of assumptions. He writes "ASSUME" on the chalkboard, then adds slashes "ASS/U/ME" while saying, "When you ASSUME—it could make an ASS out of U and ME!"

It's the hidden assumptions that can get us into so much trouble.

Consider this story. A man's son is in a terrible accident. He rushes the boy to a hospital where his son is whisked away into the operating room. The surgeon, upon seeing the boy exclaims, "Oh no! This is my son!"

If you haven't heard this one before, you may be puzzled. How could this boy be the son of the man who rushed him to the hospital and of the surgeon in the operating room?

Answer: the surgeon is the boy's mother. Now how many of us assumed the surgeon was a man?

For a long time, mathematics, the queen of all science, was assumed to be perfect. That is, it was complete and consistent. Then in 1931, Kurt Gödel came along and proved that mathematics is not complete. Gödel proved there were theorems (or statements) that could never be proved to be true or to be false unless new assumptions were brought in. But if new assumptions were brought in, then there were other theorems that couldn't be proved one way or the other.

So how did Gödel make an ass out of mathematics?

He created a mathematical statement that essentially said (after gross simplification), "This statement is false." Now clearly this sentence cannot be true or false—therefore it cannot be proven.

(See *Gödel, Escher, Bach: An Eternal Golden Braid* by Douglas R. Hofstadter for a beautiful account in layman's terms and without the gross simplification.)

Another place where we get stung by assumptions—but where we delight in being fooled—is at a magic show. Magic tricks are based on the strong human tendency to make assumptions. Let's consider an example. A magician places his beautiful assistant in a trunk and she disappears. A second later the assistant bursts out of a closet twenty feet away. The trick is, we assume she is the same person. But the truth is that she is a twin.

Obviously, rocket scientists have to be very careful about checking their assumptions. Sometimes, the mistakes they make can be catastrophic, other times just embarrassing. Remember one of the mistakes we made with Alan Shepard? We assumed that Alan wouldn't have to pee during his countdown, suborbital flight, or splashdown in the Atlantic Ocean. But this assumption ended up turning his spacesuit into an expensive diaper—and probably added a little more splash during his splashdown.

# PART V

# Simplify

"Any intelligent fool can make things bigger,
more complex, and more violent. It takes a
touch of genius—and a lot of courage—to
move in the opposite direction."

Albert Einstein

# 28

## Keep It Simple, Stupid

The 248-page *Columbia* accident report came out in August 2003, six months after the spacecraft disintegrated during reentry. Seven astronauts perished on February 1, 2003.

I read the report the day it was released. When I got to page 14, I took out my red pen and underlined the following statement: "The Shuttle is one of the most complex machines ever devised." At the top of the page I spelled out in capital letters: "KISS" and added "That's the problem!"

I first heard of the KISS principle—Keep It Simple, Stupid—in 1979 when I started working at JPL. I found it puzzling. I never met anyone at the lab that I thought was even remotely stupid. Twenty percent of the employees had Ph.D.s—nearly a thousand doctors of virtually every branch of science and engineering. There were aerospace, electrical and mechanical engineers, and astronomers, astrophysicists, and mathematicians. Every once in a while I ran into a less obvious specialist: a biologist, a philosopher, a soil mechanic.

So who were they referring to as "Stupid?" (A Navy guy told me I had it wrong—it was supposed to be "Keep It Simple, Sailor," but I am discarding this hypothesis.) The people at the lab weren't stupid. Then I heard Richard Feynman, who often gave lectures to the Caltech–JPL community, say (apparently referring to himself), "Even Nobel laureates can say and do stupid things."

What JPLers were doing was anything but simple. Dual-spin dynamics, multi-body celestial mechanics, tests of general relativity, and deep-space navigation are by their very natures exceedingly complex subjects involving advanced mathematics, millions of lines of computer code, and billion-dollar spacecraft. How can you keep it simple?

The idea (and this took me a while to get) was not to make it any more complicated than it already was. Albert Einstein put it this way: "Everything should be made as simple as possible—but no simpler."

There is a tendency for rocket scientists to sound like rocket scientists instead of human beings, analogous to students trying to sound "literary" when asked by their English teacher to write an essay. These bad writing habits seem to originate in grade school with the assignment, "How I spent my summer vacation," where the kids tell the teacher what they think she wants to hear.

The neophyte rocket scientist tries to impress his boss by demonstrating his Ph.D. prowess: gobs of equations, recondite theorems, reams of numbers. The new rocket scientist thinks his boss will appreciate this morass of technical detail. Just like the essay student who tries to please her teacher with big words and complex sentences—and without a clear message or story.

So the boss issues a caveat to all new Ph.D.s: follow the KISS principle.

There is more to this concept, however. It is a well-known principle of design that simpler systems have fewer failure modes. Simpler systems provide the anti-Murphy strategy: If there is less that can go wrong, then less will.

A simple capsule design got men to the moon and back. These capsules were robust, easy to fly, and had simple backup systems. Now consider the shuttle: It has millions of parts, it is difficult to fly, and it has few viable backup systems. It is the very antithesis of simplicity.

The KISS principle was completely ignored when NASA designed and built the shuttle.

# 29

## Draw a Picture

One way to simplify a problem is to draw a picture.

For example, say you are contemplating rearranging the furniture in your home. If you have a lot of heavy furniture and a complicated layout to your house or apartment, you can save yourself unnecessary grunt work by sketching out your floor plan and penciling in the new arrangements you are thinking about. You could take this a step further by cutting out pieces of paper that represent chairs, couches, and tables and by shifting these scraps around to see how things work out.

If you have drawn an accurate representation, then you can determine—without ever leaving your armchair—which possible arrangement makes sense.

Rocket scientists call such drawings blueprints. They used to draw blueprints by hand, which was a tedious, painstaking task because standard practice required accuracy to within $\frac{1}{32}$ of an inch. Drafting courses were taught to generations of engineers. I took such a course and hated every minute of it. My impression was, "Here's a hundred-year-old course being taught by a hundred-year-old man." Three views (front, top, and side) which provided three-dimensional information had to be drawn precisely, and labeling had to be done in rigorous block letter form.

Now most drafting is done on computers, consequently the ability of rocket scientists to draw or print legibly has atrophied. But we still use the phrase, "Back to the drawing board" whenever we get bit by Murphy's law and have to redesign a spacecraft or rocket.

It is unfortunate that most rocket scientists can't draw. I think that, in addition to learning computer graphics, students of rocket science should take a course on drawing from the art department. Because we are visual creatures, the ability to make a credible drawing on a piece of paper, a napkin, or a chalkboard is a valuable aid to thinking. Most engineers are visualizers and need pictures to understand the problem they're working on.

In many ways, drawing a picture is a method of simulating and solving a problem. When you sketch your floor plan and consider various furniture arrangements, you are making an analog computation—a simulation. You don't know how things will turn out until you've drawn a picture—and suddenly you see the solution. You've solved the problem, actually computed it, without using any mathematics!

# 30

## Make a Mock-up

Rocket scientists, after they are finished dreaming, go through three stages called design, build, and test. The design may start innocuously with a sketch on a paper napkin during lunch. Before actually building a spacecraft, more accurate drawings, namely blueprints, must be made.

Of course, a great deal of analysis and simulation must be done first so that the hardware depicted in the blueprint accurately reflects the capabilities the spacecraft must have to achieve its mission. Spacecraft accoutrements typically include thrusters, propellant tanks, sun and star sensors, gyros, accelerometers, antennas, computers, solar arrays or nuclear batteries, and a host of other navigational, life-support, and scientific equipment. Sometimes, engineers get so involved with these details that they forget to look at the big picture (the blueprint) to see how it's all put together.

I have noticed this tendency in my senior design students. Students are reluctant to draw a picture or create a CAD (computer-aided design) drawing of their spacecraft. They become mired in a sea of calculations on orbital mechanics, material strength, power requirements, telecommunications, and so on. And yet it is the picture that makes the design credible (assuming that all the relevant hardware is depicted and that the calculations are correct). A picture with dimensions and locations of all the major subsystems demonstrates the feasibility of the design. Nearly every piece of equipment requires power, occupies volume, and adds to the total mass (or weight). And mass on a spacecraft means money. The problem rocket scientists face is: How do I fit all the stuff I need into this spacecraft—and within the weight limit?

Let's say you are a backpacker about to hike down to the bottom of the Grand Canyon and camp for a few days. You ask yourself three questions. How much stuff do I need? Will it all fit in my backpack? Will it be too heavy? The answers turn out: lots, no, and yes! Now you can start thinking like a rocket scientist by unpacking and repacking. You examine each piece of gear and eliminate the dispensable. You consider how one piece of equipment can serve multiple purposes. Do you really need to carry a fork, a knife, and a spoon? Or will a spoon serve all three functions in a pinch? These kinds of questions can lead to new inventions or ways of doing things—like the spork, a combination spoon and fork. (It sounds dopey, but it works.)

When you go through this process of packing and repacking your backpack, you are performing a type of simulation that rocket scientists call a mock-up. A set of blueprints is usually not enough to answer all the questions that come up in spacecraft design. Rocket scientists often build a three-dimensional mock-up. It may be a small-scale model at first, followed by a full-scale model. For piloted missions, full-scale mock-ups are requisite. In the movie *Apollo 13*, the need for a mock-up is dramatically demonstrated as astronauts on the ground test out emergency procedures before implementing them on the real spacecraft. Three-dimensional mock-ups tell us if all the equipment fits, if the instruments and controls are well placed, if the astronauts are able to perform their mission.

Artists and architects use scale models for similar reasons. The nineteen-foot statue of Abraham Lincoln that sits in the Lincoln Memorial was designed by Daniel Chester French first as a small-scale model, then a larger model, and finally the real thing. The 151-foot Statue of Liberty was scaled up from the original four-foot model designed by the French artist Bartholdi. The index finger on the actual statue is eight feet high, and the total weight of the statue is 225 tons. It is inconceivable that a sculpture of this magnitude could have been built without the use of scale models.

Not all mock-ups are sophisticated or precise. As mentioned before, rocket scientists have used Tinker Toys to build simple mock-ups of spacecraft so they could easily visualize the vehicle's configuration and motion in three dimensions.

So, after drawing pictures, rocket scientists often turn to three-dimensional representations. These mock-ups are used to simulate and solve design problems. Whether they are constructed out of virtual images, plywood, plaster, or Tinker Toys, mock-ups are a crucial thinking tool in the rocket scientist's arsenal.

# 31

## Name the Beasts

A powerful method of simplifying is to make up "handles" for new problems; that is, to create a nomenclature. Human beings are the world champions at doing this—inventing language. In fact, every individual human being has the capacity to create language.

An experiment showed that people will invent terminology spontaneously. Two people were placed in separate rooms where they could not see each other but could communicate via intercom. Each subject was given a sheet of paper with sixteen pictures on it, arranged in a four by four grid. The same pictures appeared on both sheets, but they appeared in different boxes on the grid. (For example, the picture in the upper right-hand corner of one person's sheet was located in a different place on the other person's sheet.)

The pictures were simple modern art sketches that had no recognizable objects in them: squiggly lines, pointed star shapes, contorted geometric shapes.

The subjects were asked to match up the pictures on the sheets with each other by talking over the intercom. Typically, the subjects started out using a lengthy descriptive phrase, "Looks like squiggly lines—waves on a beach." Later the phrase was shortened to "squiggly lines." Finally the abbreviated form, "squiggle," appeared and was quickly adopted for the remainder of the discussion.

How easy and natural it is for human beings to invent terminology!

Far simpler than using lengthy descriptions.

Of course, rocket scientists have had to invent novel terms for the new technologies they were dealing with. They took it a step further with their ubiquitous use of acronyms. So besides truncated terms like "capcom" for capsule communicator and "retro" for retro-fire engineer, we have "FIDO" for flight dynamics officer and "NASA" for National Aeronautics and Space Administration.

New words, new terminologies, new problems. We use words to categorize, to classify, to put things in order. In our increasingly complex world, things can seem chaotic and random at times. Human beings create order out of chaos by naming things to "get a handle on it." Our use of language is our greatest survival skill.

Because of the power of language, great care must be exercised in its use. Merely labeling something does not mean we understand it. We can misuse language in many ways. We can oversimplify, misconstrue, stereotype, malign, and otherwise misspeak. We have to be careful how we name things—we must endeavor to be precise.

A beautiful example of losing sight of precision in language was described in an article in *Scientific American* about a word game called "Tower of Babel." In this game, you start with a word and look up a synonym in a thesaurus. Let's say we start with "disrespect" and find "disregard." Next we look for a synonym of disregard and find "allow." In a few steps, we can often find a word that is the opposite—an antonym—for the first word. Here is my example:

> disrespect
> disregard
> allow
> approve
> commend
> praise
> revere
> respect

I was astounded by the idea. I had always loosely assumed that synonyms were equal like the symbols in an equation. If A equals B and B equals C, then A equals C. Carried to an extreme, A equals Z. But the "Tower of Babel" proved this assumption to be incorrect. (Be careful what you assume.) Synonyms are more like the colors of the rainbow, which exist on a continuous spectrum from red to violet. (See A. K. Dewdney's description of Ron Hardin's game in "Word ladders and a tower of Babel lead to computational heights defying assault" in the August 1987 issue of *Scientific American*. Hardin unearthed many interesting short chains such as: *acceptable* → *so-so* → *ordinary* → *inferior* → *rotten* → *unacceptable*.)

We should choose judiciously from this spectrum of words (this slippery slope of synonyms) and endeavor to avoid slanting our choice to either side. We need to watch our language when naming the beasts.

# 32

## Look at the Little Picture

The preferred method of virtually all science is to break a problem down to its simplest component parts: to look at the little picture. Although sometimes referred to disparagingly as "reductionism," there can be little doubt of its success or power.

In the delightful movie *Creator*, Peter O'Toole plays a Nobel prize–winning professor who carries his deceased wife's cells around in a thermos bottle in hopes that he can someday re-create her. He's a charming eccentric who reminds everyone he talks to that he is "looking for the Big Picture." At one point, he puts his philosophy into the most dramatic metaphor he can think of: "I want to know what God's testicles are doing!"

His rival colleague, another professor, counters with, "There's no such thing as the Big Picture. There's just a bunch of little ones."

The rocket scientist, however, realizes that both modes of thinking are crucial to the success of a space mission.

Henry Ford said that every problem, however complex, can be broken down into a series of simple steps. He proved his vision by developing the assembly line and putting America on wheels. Ford's divide-and-conquer scheme made it possible to build automobiles by the million. In *Rocket Man*, David Clary reports that Robert Goddard had boiled down space travel to just twenty-six steps. (Clary tells the story of America's first rocket man brilliantly, humorously, and entertainingly.)

The idea of reductionism goes back to the ancient Greeks. In the fifth century B.C., Democritus thought about cutting an object into halves, then cutting the halves in half, and continuing the process until he obtained particles that could not be cut in two. He named these particles "atoms," which means "indivisibles." Democritus believed that our entire universe was built up by the collection of a vast number of atoms and the void between them.

There is an old story of six blind men who examine an elephant. The first man announced, after touching the elephant's side, "The elephant is like a wall." The second blind man, who had been examining the elephant's tusk, declared, "The elephant is like a spear." The blind man at the elephant's trunk determined, "The elephant is like a snake," while the man at the elephant's knee proclaimed, "The elephant is like a tree." The fifth man, feeling the elephant's ear concluded, "The elephant is like a fan." The sixth at the elephant's tail said, "The elephant is like a rope." All of the blind men were right about certain aspects of the elephant. But none of them had the Big Picture. (How could they? They were all blind.) The moral of this story is, of course, a warning about reductionism. None of the blind men had the whole elephant. And everyone knows the elephant is more than the sum of its parts.

Rocket scientists understand this caveat. They must deal with a myriad of little pictures. Each of the component parts must work so that the little pictures will add up to the Big Picture—the exploration of space.

# 33

## Do the Math

If you can convert your problem into a math problem, then you have done two things: (1) you have simplified the problem and (2) you have brought the full power of mathematics to bear on your problem. Translating a problem into a mathematical one is called mathematical modeling. Once a math model is chosen (which means that certain specific assumptions are made), there are centuries of mathematical techniques that can be used to answer your question. In this way, you tap the brains of mathematicians who have already solved your problem.

Unfortunately, many students have been permanently turned away from the power of mathematics by the "story problems" they were forced to solve (or at least forced to struggle with) in grade school.

"A train leaves San Francisco traveling south at 45 miles per hour. Meanwhile, another train leaves Los Angeles traveling north at 50 miles per hour. A fly, traveling at 100 miles per hour flies from the first train to the second train and continues back and forth until the trains collide. Assuming San Francisco and Los Angeles are 380 miles apart—how far does the fly fly?"

If this kind of problem makes your head spin, or makes you feel like you're about to go crazy, or simply breaks your heart—then you are not alone.

There were at least good intentions on the part of the math teachers who introduced story problems that had to be converted to math (arithmetic, algebra, or geometry). But they picked such awful problems. Who cares how far the fly flies? The teachers forgot about *ganas*—desire. Students must have a strong reason for solving the problem.

Albert Einstein was the world's greatest scientist, but he did not care about calculating numbers for the sake of exercising his mathematical dexterity. Actually, Einstein was only an average mathematician. He often made simple calculation errors, some appearing in his published works. Einstein said he did not care about how a particular atom vibrated. He wanted to understand how the universe was constructed. His Big Picture approach was so driven by his desire to know, that he slogged through the horrible complexities of tensor calculus.

Einstein viewed math as a necessary tool for communicating his findings to other scientists. In fact, Einstein explains that he used his physical intuition to solve problems—a combination of visual imagery and muscular feelings. In a visceral way, he knew when he had the solution. Then he would have to prove the results of his thought experiments to the scientific community by writing down the attendant mathematics. Einstein's advice to the rest of us is "Do not worry about your difficulties with mathematics; I can assure you that mine are still greater."

If your problem can be written as a mathematical statement, then you have a great advantage in solving your problem because there are vast mathematical resources at your disposal.

Rocket scientists find math indispensable for calculating the flight of a spacecraft to another planet. Not all rocket scientists are great mathematicians, but they all have a great desire to make their mission succeed. Even Einstein did not know about tensor calculus when he started his quest for the general theory of relativity.

Mathematics may not be your cup of tea, and this may be due in part to painful experiences in school. Mr. Escalante demonstrated that he could teach advanced calculus to underprivileged high school students in East Los Angeles. He instilled in his students a great desire to learn, and they proved themselves brilliantly on the standard placement tests.

Math is not the problem—lack of desire is.

(By the way—if you're still curious—the fly flies four hundred miles before the trains collide.)

# 34

## Apply Occam's Razor

*Occam's razor* states that the simplest explanation is probably the correct one.

Suppose you hear noises in the middle of the night, and the next morning you discover a broken lamp in your living room. You can construct a number of hypotheses to explain the events. Alien beings, from a small planet in the Orion Nebula, have landed their flying saucer in your backyard, tiptoed into your living room, and just before they were about to abduct you and perform invasive biological experiments on you, they tripped over the lamp and were frightened off.

That's one hypothesis.

Consider another. Last night you forgot to let out your cat, Snowball 3. After lights out, your dog, Rover 5, who usually sleeps quietly in the living room, chased Snowball 3 who knocked over the lamp while scrambling to get away.

Obviously, the simplest hypothesis is the second one. A version of William of Occam's axiom (from *Bartlett's Familiar Quotations*) is, "Entities should not be multiplied unnecessarily." We already have two entities (Snowball 3 and Rover 5)—do we need a third?

By checking to see if your cat is still in the house, you can confirm a prediction of this hypothesis. Of course, if there are three-toed footprints, no Snowball 3 to be found, and a huge depression in your backyard where a heavy object rested during the night, you might reconsider the first hypothesis. The simplest explanation is probably correct, but not always.

Did the United States fake landing on the moon? There is a hypothesis that NASA discovered that it was impossible to get to the moon on President Kennedy's timetable, so they went to Hollywood and faked everything with special effects.

The alternate hypothesis is that we really did land a man on the moon. On July 16, 1969, around a million people went to Cape Canaveral and witnessed the launching of the gigantic moon rocket, the *Saturn V*. It stood 365 feet high, as tall as a 36-story building. It shook the ground when it took off. (Kurt Vonnegut's brother said the noise was so loud even at three miles away that he was convinced it was worth every penny of the $30 billion we spent.) Those million people saw it ascend into the sky, pick up speed, drop off the first stage, accelerate, shrink, and finally disappear. Ham radio operators from around the world listened to the astronauts, Neil Armstrong, Buzz Aldrin, and Michael Collins communicating with Mission Control. Lunar rocks—unlike anything on Earth—were brought back along with many photographs. More than 400,000 people worked on the project for nearly ten years. Three men died (Gus Grissom, Ed White, and Roger Chaffee) in a horrible fire during a ground test of the *Apollo* capsule. Instruments were placed by Armstrong and Aldrin that allowed laser beams from Earth to be reflected off the moon to determine the distance within centimeters. A dozen men walked on the moon, returned to Earth, and told their stories.

The alternative hypothesis would suggest that the giant moon rocket wasn't going to the moon (or at least did not carry men there), that three men were killed to make the story seem real, that hundreds of thousands of scientists, engineers, technicians, and skilled laborers kept the secret along with the dozen men who claim they walked on the lunar surface.

As Sherlock Holmes told Dr. Watson, "When you eliminate the impossible—whatever remains—however improbable, must be true."

I submit that it would be impossible to keep the secret of faking the moon landing when so many people were involved. In fact, it would be a lot easier to actually go to the moon than to fake it.

# PART VI

# Optimize

"If a man can write a better book, preach a
better sermon, or make a better mousetrap
than his neighbor, although he builds his
house in the woods the world will make a
beaten path to his door."

Ralph Waldo Emerson

# 35

## Minimize the Cost

Everyone wants to save a buck. In this chapter and the next, I diverge from the ordinary-life lesson and talk about how rocket scientists try to save time and money.

When a shuttle astronaut drinks a sixteen-ounce bottle of water it costs about $10,000.

Why is it so expensive? There are two reasons. The first is the poor design of the shuttle itself, which has unfortunately made space travel more expensive (not to mention more dangerous) than it has to be. (More on this point later.)

The second reason has to do with the laws of orbital mechanics. To get that bottle of water into orbit (along with the astronauts, their life-support system, and everything else), it has to be accelerated to a speed of five miles per second. After the rocket engine burns out, the bottle will be in free fall—it will fall toward Earth, but it is moving so fast that Earth's surface curves away from the bottle at the same rate that the bottle falls toward the ground. These two effects, the falling of the bottle and the curving of Earth's surface, cancel out at five miles per second. To be placed in a circular orbit, the bottle must be moving parallel to Earth's surface at this incredible speed.

To make this concept of circular orbit clearer, let's do a thought experiment. Imagine driving your car off a cliff over the Grand Canyon. (And please remember that this is only a thought experiment.) If you drive at sixty miles per hour, your car will continue at that speed after your wheels leave the road and you start falling into the canyon. (We are neglecting the effect of air drag, which makes the problem a bit more complicated but doesn't change the basic concept.) How long it takes you to hit the ground does not depend on your speed (of sixty mph) but only on the constant acceleration (i.e., the pull) of gravity and your height above the canyon floor. If you drove at 120 mph, you'd still hit the ground at the same time, but you'd be twice as far down range. The faster you go, the farther down range you will travel, always hitting the ground at the same time. This analysis starts to break down when you travel at very high speeds, because you cover so much down-range distance that the curvature of Earth starts to matter. At five miles per second, the surface of Earth curves downward at exactly the same rate that you accelerate downward from the pull of gravity. Your car never hits the ground because the ground is falling below you due to the curvature of our planet. While you and your car fall together, you feel the sensation of falling. Because your car falls with you, you seem to be floating inside the car. You feel you will crash into the ground, which is exactly how astronauts in orbit feel: You are experiencing weightlessness. (It takes even astronauts a while to learn to ignore the sensation that they are falling to their deaths. After a few days, the vomiting stops and it can become a pleasant experience.) In this thought experiment, we have ignored the effect of drag, which would slow you down and cause your orbit to decay. To circumvent the orbit decay problem, spacecraft have to be at least one hundred miles above Earth's surface where the atmospheric drag is negligible.

So weightlessness is not the absence of gravity at all—weight-lessness is falling under the pull of gravity. If you want the sensation to last indefinitely (and who doesn't?), then you have to be traveling at the fantastically fast speed of five miles per second. (This is technically known as the "circular speed" for low-Earth orbit.) At this speed, a spacecraft will zip past the entire United States in under eight minutes; in ten minutes it will cross the Atlantic Ocean.

To get the spacecraft into orbit, it is placed in the payload bay of a launch vehicle such as a *Delta*, an *Atlas*, or the shuttle. The launch vehicle is ignited and launched straight up into the air. As it ascends vertically to greater and greater heights, it starts to tilt over to one side. This tipping is intentional. The rocket continues to tip while gaining altitude until it is no longer traveling vertically, but instead is traveling horizontally. When the rocket reaches a height of one hundred miles it achieves a speed of five miles per second and is traveling parallel to Earth's surface. The engine cuts off, and the astronauts experience falling without ever hitting the ground (i.e., weightlessness).

The tipping rate of the launch vehicle is determined by a control scheme called "the steering law." If the tipping occurs too fast, it will cost more pounds of propellant to get into orbit. If it occurs too slowly, it will also cost more.

Rocket scientists use a type of mathematics called "calculus of variations" to find the cheapest way to get into orbit. According to the theory, the steering law that takes the minimum time to reach orbit is best because the rocket burns the least amount of propellant.

When rocket scientists apply this technique to the shuttle, the result is that it costs $10,000 to give an astronaut a drink of water!

# 36

## Minimize the Time

The concept behind optimal space trajectories (e.g. the cheapest way to get into orbit) came to us from the Bernoulli brothers. In the seventeenth century, James and John Bernoulli amused and challenged each other by inventing mathematical puzzles for the other to solve. It was typical sibling rivalry carried to extremes. The game reached its zenith when one of the brothers proposed what is now called the "Brachistochrone problem." (Consider it the brontosaurus of mathematics.) The name derives from the Greek roots that appear in brachiopod and chronology, so it translates to the "shortest time" problem.

The problem is very similar to that of launching a spacecraft into orbit. Because the propellant is burning at a furious rate, we burn the least if we get into orbit by the quickest path.

To understand the Bernoulli problem, let us think of a roller coaster. Suppose we have a coaster with a hundred-foot drop over a hundred-foot range. What shape would the track have to be so that the roller coaster gets to the bottom in the shortest time? This is the Brachistochrone problem. Suppose we build a straight-line track—that is, an inclined plane—that runs from one hundred feet high down to the bottom, one hundred feet along the ground. It would be on a 45-degree angle. Wouldn't this be the fastest path?

The answer, surprisingly, is no. It's the shortest length but not the fastest path. (And these people want to get down there as quickly as possible—they want some thrills in their lives.) It turns out that you can satisfy your thrill seekers and get them to the bottom faster if you have the track drop more steeply at first. It will seem to your riders that they are dropping almost straight down on the first plunge. (Serves them right—these people really bring out the sadist in the amusement-ride designer.) A steeper drop in the beginning gives them greater speed in the beginning (gets their hearts pumping) and this extra speed more than makes up for the longer track (compared with the soporific straight-line incline). Make it too steep, however, and the total trip time gets to be longer (and therefore boring).

The problem the Bernoulli brothers invented stumped the great mathematicians of Europe for six months. (The problem is to find a mathematical function to describe the exact shape.) A wonderful account is given in Eric Bell's *Men* [sic] *of Mathematics* (which does discuss some female mathematicians) and is an inspiring read. Even Isaac Newton worked on the problem, but unlike everyone else, he solved it in a single day. Both James and John came up with their own solutions.

Then an amazing thing happened. The scientists of Europe discovered that the equation that solved the Brachistochrone problem is the same equation that governs the basic laws of physics—the laws of motion, of the planets, of everything!

It seemed that Mother Nature was an optimizer herself. Nature finds the shortest time path for a beam of light traveling through air, water, and glass. Because light travels slower in glass, it avoids spending too much time in the glass and takes a more distant path through the air where it travels faster (than it does in glass). This is Fermat's principle of optics. It is the same strategy a life guard follows to save a drowning man. If the drowning man is located on a diagonal path across the water, she will run as fast as she can along the beach and only enters the water at a point where she will get to the drowning man the quickest. She knows that she swims a lot slower than she can run. She is an optimizer.

The principle of the shortest time (and related optimization problems) is ubiquitous in nature and in human behavior. It applies to quantum physics, general relativity, optics, space trajectories, traffic control, aircraft design, game theory, and population dynamics. All of these theories and applications came from the original work of the Bernoullis who fought like brothers but thought like rocket scientists.

So how can we apply the principle to ordinary life? Perhaps Benjamin Franklin said it best in his advice to a young tradesman: "Remember that time is money."

# 37

## Be Mr. Spock

Mr. Spock of the original TV series *Star Trek* is the classic Analyst. He is driven by logic and firmly believes—actually he knows—there is one best way to do things. In *The Art of Thinking*, Allen Harrison and Robert Bramson describe the Analyst as one of the five styles of thinking. (Ranging from right brain to left, these are personified as the Synthesist, the Idealist, the Pragmatist, the Analyst, and the Realist.) But Gene Roddenberry depicted the Analyst earlier—in the character of Mr. Spock. Spock was brought to life by the acute interpretation of Leonard Nimoy who wrote two memoirs about his experience that seem to highlight his ambivalence about the role, perhaps due to the internal schism of the character: *I Am Not Spock* and *I Am Spock*.

Throughout the *Star Trek* series there is a constant struggle between Captain Kirk's closest advisors. Mr. Spock presents the analytical, logical thing to do—which sometimes appears to be heartless and cruel, even shocking. Dr. McCoy (played with embarrassing realism and humanness by the late, great DeForest Kelley) insists on the human thing to do. The character of Dr. McCoy matches remarkably well with the Idealist of Harrison and Bramson. The Idealist has high ethical standards, elevates people over facts, focuses on a holistic approach, and trusts his intuition over logic or "the one best way."

Debates rage between Spock and McCoy who try to sway Captain Kirk to their way of solving a crisis. Often the suggestions they make to Kirk are unacceptable to him, and he demands a third alternative to resolve the dilemma he faces. Interestingly, Kirk is the Pragmatist. He is humorous, personal, enthusiastic, tactical, and sometimes appears insincere. Where McCoy may be a "Bleeding Heart" and Spock a "Great Stone Face," Kirk is a "Politician."

But *Star Trek* fans (and some of them are rocket scientists) know that Mr. Spock is the real hero of *Star Trek*. There are times when the Analyst must prevail. There are times when you want the one, tried-and-true, best-and-only way to do something.

During brain surgery for example. You don't want creativity—you want perfection.

Often rocket scientists are thought to be pure Analysts, even the epitome of what Analyst means. And certainly there are times to emulate the Analyst—to be Mr. Spock.

But as we have seen throughout this book, the rocket scientist must draw from the full spectrum of the styles of thinking, so well described by Harrison and Bramson.

# 38

## Make It Faster, Better, Cheaper (But Not All Three!)

When Dan Goldin became NASA administrator in 1992, NASA was facing a crisis: cut costs or go out of existence. It was under this extreme pressure that Goldin hatched his slogan, "Faster, Better, Cheaper," and sold it to Congress. Goldin's idea was born out of self-preservation; fierce political winds were blowing and only the strong would survive. Goldin came up with his three-pronged optimal solution: NASA would be "the best" by all three measures.

But as Cliff Stoll said in his enlightening book *High-Tech Heretic*, you can't have all three. (Stoll was the astronomer who broke up an international spy ring by noticing an excess charge of seventy-five cents on his Harvard main frame computer; he tells his spy-thriller, scientist–detective story in his gripping memoir, *Cuckoo's Egg*.)

Let's say you're hungry and you want to eat dinner. As Stoll points out, you have three choices: you can have it better and faster by going to an expensive restaurant, you can have it better and cheaper by taking the time to cook it at home, or you can have it faster and cheaper by settling for a burger at McDonald's. You have to give up one of the three: economy, speed, or quality. But Dan Goldin was no dummy—he understood that he could pressure NASA to really look at this optimization trade.

There is a branch of optimization theory that considers multiple goals that are contradictory. We face such problems all the time in real life. In the *Galileo* project, I had to design a sequence of eleven orbits about the planet Jupiter that would provide the principal investigators with observational opportunities to "maximize science return." Unfortunately, there were three major objectives: study the moons of Jupiter, study the atmosphere of Jupiter, and study the magnetic field of Jupiter. And guess what? These goals were at odds with each other.

As a mission designer, I found myself in a role similar to a AAA travel agent trying to satisfy a family of three—each of whom wants to visit a different sight. What makes their problem difficult is that they're on a tight schedule and budget and have only one car to drive. The travel agent can only draw one Trip-tik—only one path that goes from A to B to C. Similarly, maximizing Jovian atmospheric observations would reduce opportunities to photograph the moons or to study the magnetosphere of Jupiter. To maximize each scientific investigation, we really needed three separate spacecraft, but we only had one spacecraft. And planetary scientists (the principal investigators) were a lot like backseat drivers on a family vacation—they weren't above throwing temper tantrums to get their way.

In the end, mission designers (in their cool, logical, Spock-like way) had to compromise and so did the scientists (though there was some screaming and shouting and shedding of tears). A trajectory (an orbital tour) was designed that would get "adequate" coverage for all major science objectives. The idea was to plan a feasible mission that would not give inadequate return to any major area.

So when Dan Goldin cooked up his shibboleth of "Faster, Better, Cheaper," he knew what he was doing. His statement is illogical, taken literally, but behind those three words there is a compromise and that was the only way NASA could survive the government budget cuts at that time.

In real life, we deal with these seeming paradoxes frequently. It is helpful to get specific about what your goals are (such as faster, better, cheaper), but you also need the wisdom to decide what combination of these goals will satisfy you. Life is a compromise (as is rocket science).

# 39

## Know When Bigger Is Better

Sometimes bigger is better. The economy size is cheaper by the pound, the gallon, the dozen. We all look for bargains in volume.

The lesson that NASA should have learned from the Soviet space program is that bigger boosters are better. When the Russians launched Sputnik in 1957, Americans were shocked into a space race in which their very survival seemed to be at stake. But how could Americans compete? The U.S. military–industrial complex had been frantically building nuclear warheads and intercontinental ballistic missiles (ICBMs) to deliver these warheads to targets in the Soviet Union.

Americans also feared that the Soviets might attempt a first-strike sneak attack to destroy all our missiles on the ground before we could launch them. So the United States built compact, highly efficient boosters that were small enough to tuck neatly away inside concrete bunkers where they could survive a nuclear blast. Then our missiles could still be launched in retaliation to a first-strike attack. This approach precipitated the mutually assured destruction policy, aptly named MAD. To make sure our rockets were small enough to fit into the concrete silos, the missiles were made of exotic, expensive materials (like high-strength, low-weight titanium) and were assembled by engineers wearing surgical gowns. Clean rooms were necessary to avoid a speck of dust jamming the complex machinery.

When the Soviets started launching men into orbit, the United States strove to catch up by putting our astronauts into tiny capsules atop ICBMs—the only launch vehicles we had. The boosters were expensive and the capsules were puny. Meanwhile, the Russians realized that bigger was better. Their launch vehicles were built out of steel, assembled by factory workers. They didn't use exotic materials, and the vehicles were inefficient—they went for size, which more than made up for the inefficiencies. Their first satellite, *Sputnik I*, was a cannonball weighing 184 pounds. The first American satellite, *Explorer I*, launched in 1958, weighed only 30 pounds. *Sputnik II* weighed a ton and carried a dog. The Soviets were able to launch huge spacecraft into orbit—soon they were launching crews of two and three men.

Americans were losing the race. They were forced to deal with cramped spaces; forced into developing miniature electronics and using astronauts "with compact builds" who were under six feet tall. Gus Grissom, one of the original *Mercury*-seven astronauts, at five-foot-seven was the perfect size; his fellow astronauts called the *Mercury* capsule the "Gus mobile." The pressure to invent microelectronics advanced computer technology. The small-is-beautiful philosophy eventually gave American electronics an advantage. Not that it was planned that way. In 1968, Arthur Schnitt, an engineer at the Aerospace Corporation (a think tank for the U.S. Air Force), proposed using large boosters that would capitalize on the economy of scale. Gregg Easterbrook tells the story in his *Newsweek* article, "Big Dumb Rockets," which appeared on August 17, 1987. Schnitt's idea was very similar to the Russian approach: it would make space travel a lot less expensive. In 1968, his classified work was reviewed and promptly canceled. The successor to the moon project would not be Big Dumb Rockets, but the highly complex and expensive shuttle.

On November 9, 1967, the first *Saturn V* was launched in an unmanned test. It was America's first heavy lifter. It was big, but not so dumb. It lifted over one hundred tons into Earth orbit. It performed flawlessly. America was on its way to the moon.

Today, because of a decision made in 1968, we have the shuttle. It is complicated, unsafe, and inefficient. It is as big as the moon rocket but it carries less than thirty tons into orbit (less than one third of what the *Saturn V* carried). The shuttle's payload is not big, nor is the shuttle dumb or cheap.

After two horrible failures of the shuttle, which cost the lives of fourteen astronauts, it is time to learn from our mistakes and build a new launch vehicle that is bigger and better.

# 40

## Let Form Follow Function

The great American architect Louis Henri Sullivan (who influenced Frank Lloyd Wright) developed the concept that "form ever follows function." This optimization principle stresses the primary goal of any design: the device should—first of all—work. Sullivan's concept also unites the operational with the aesthetic; after all, he was designing buildings that were both efficient and beautiful.

Although much has been written about the meaning of "form follows function," we will take the simple interpretation of "don't put the cart before the horse."

In the *Mercury*, *Gemini*, and *Apollo* programs, NASA took a completely operational, logical approach to getting a man on the moon. They looked at the major obstacles: surviving in space, walking in space, and rendezvousing in space. The one-man *Mercury* capsule tested whether a human being could even survive in weightlessness. Some experts thought you would choke to death if you tried to eat when there was no gravity to move food through the digestive tract. The two-man *Gemini* program proved that astronauts could spacewalk—allowing transfer of a crew from one vehicle to another if docking the two craft together ever failed. *Gemini* proved that rendezvous of two vehicles traveling at 17,500 miles per hour around Earth was possible—it paved the way to lunar rendezvous. The *Apollo* program put it all together. Astronauts could live in space long enough to travel to the moon and back. Lunar rendezvous enabled a smaller craft to land on the lunar surface and to return to the orbiting mother ship.

The shapes of the space vehicles derived from their functions, as Louis Henri Sullivan would have it. Blunt capsules for reentry into Earth's atmosphere; a spindly spider-like vehicle for landing on the airless moon. The problem of landing a man on the moon and bringing him back safely was approached by looking at the final goal and working backward through all the smaller steps that had to be accomplished.

Now let's contrast the highly successful moon project with the subsequent situation at NASA.

At one time there was a plan, before the completion of the *Apollo* program, to establish a lunar base that would be a stepping stone to Mars exploration. The lunar base would be supported by a space station in orbit around Earth, which would be supported by a shuttle. In Stanley Kubrick's classic film *2001: A Space Odyssey*, we see a beautiful representation. A sleek Pan American shuttle takes Dr. Haywood Floyd up to the rotating space station, a wheel in the sky, and docks majestically to the tune of the *Blue Danube*; then a nonaerodynamic lunar transfer vehicle, an immense sphere, takes the good doctor to the lunar base. Stanley Kubrick was not a rocket scientist, but he had the advice of Dr. Arthur C. Clarke, who was. Their picture is not bad at all—it adds up to a credible depiction of how it could actually be done.

But in the 1970s, President Richard Nixon put the kibosh on the shuttle–station–moonbase program. He canceled the *Apollo* program (after making his historic phone call to Neil Armstrong during the first moon walk), canceled the space station, but kept the shuttle. Mr. Nixon's decision has hobbled our space program ever since. Without a space station, the shuttle had nowhere to go. And now that we are in the middle of building a space station, we have no lunar base that it would support.

Our politicians have seized the form of space exploration, but not the function. Congress funded the shuttle with no clear purpose. It is funding the space station with no clear future. The shuttle itself is a classic case of breaking the form-follows-function rule. Here we have an airplane (a glider really) that carries wings into space—at the cost of $10,000 per pound. It lands on a runway with the "dignity" of a high-performance aircraft. Pilot–astronauts seem to enjoy the pomp and circumstance of flying this winged-and-wheeled craft. Some of these pilot–astronauts might dismiss the Spam-in-the-can capsules of the *Apollo* days. But why should we care what the style of the vehicle is? It is the mission that matters—not the looks of the vehicle.

When Armstrong, Aldrin, and Collins returned to Earth from the moon in their cramped capsule—which landed in the Atlantic Ocean under the canopy of three parachutes—no one complained about the indignity of this ending. They had participated in the first human landing on the moon! It was the mission that mattered.

The trouble with NASA's shuttle program is that it had no clear mission. Unless we set a clear goal, Louis Henri Sullivan's form-follows-function guideline cannot help us.

# 41

## Pick the Best People

Sometimes problems are solved by assembling a team. Aerospace companies, graduate schools, and government think tanks often rely on the ability of a group of experts to crack a tough problem.

There are, of course, advantages and disadvantages to the two-heads-are-better-than-one approach. The most important team-building caveat is expressed by Jeffrey Fox in his insightful book *How to Become a Great Boss*. Fox scores a weak or average manager a 7 on his 10-point scale of competency and then observes that 9s pick 9s and 10s, but 7s pick 5s, and 5s pick 3s. The most qualified, confident team leaders select people of equal or greater talent compared with themselves. They are unafraid of brilliant, creative people and look forward to working with them and learning from them. Weaker leaders fear smart workers, so they hire individuals of lesser skills so as to not feel threatened. This tendency of a weak leader to fear extraordinary ability in others will result in the creation of a weak team.

Fox's advice is clear: Pick the best people, starting with the managers.

In the best book written (so far) about human space exploration, *The Case for Mars*, Dr. Robert Zubrin talks about the composition of the first crew to land on Mars. He suggests that there will be no need for a Captain Kirk or a Dr. McCoy. Instead, for a crew of four it would be best to have two Mr. Spocks and two Scottys.

The need for a strong commander to captain the ship and issue orders will not be as important as it is presented in *Star Trek*. The spaceship to Mars will be highly automated and therefore will not require a highly trained pilot to fly it. The crew will be exceptionally intelligent and motivated to begin with—and will not need a disciplinarian to crack the whip. A medical doctor who has spent a great deal of time training to recognize and treat a wide variety of traumas and diseases will not be required. The crew will be excellent specimens of health, screened for all possible illness and medical problems. In fact, astronaut medical screenings are so thorough (some would say invasive—they examine every nook and cranny) that often a candidate will learn for the first time that he or she has an eye defect, a kidney stone, or even a tumor.

The crew will be trained to perform first aid and certain paramedical procedures. They will not have to treat each other for heart disease, colon cancer, diabetes, AIDS, or Alzheimer's. They won't have to perform hysterectomies, colostomies, open-heart bypass operations, or brain surgery. They won't have to constantly examine and treat the captain for rare STDs picked up from alien encounters of the fourth kind (though this may have been Dr. McCoy's main function).

Much more important in a human mission to Mars is the ability to get there and to get back and the reason for going there in the first place (science and exploration). Two Mr. Spocks will serve science well. Two Scottys will provide the engineering to maintain and repair the ship to ensure a safe return.

Too bad for Captain Kirk and Dr. McCoy. But their fate was already predicted in the *Star Trek* episode "The Ultimate Computer." They were designated "nonessential" personnel.

# 42

## Make Small Improvements

One approach to finding the best solution to a problem is to start with a good solution and to then improve upon it. As World Chess Champion Dr. Emanuel Lasker said, "When you find a good move, look for a better one." The great success of the Volkswagen Beetle was based on this concept. Every year small changes were made—only those that improved the vehicle. For decades, "the people's car" was the most popular in the world.

The Boeing 700 series aircraft follow a similar theme. The Boeing 707 established the archetype for the later models. This nearly optimal design spawned larger, more efficient versions: the 727, 737, 757, and the 747 jumbo jet. The reason these aircraft look so similar is that they are so close to the perfect solution of what a subsonic jet airliner should be.

Why are they called the 700 series? Because Americans believe in lucky numbers and "7" is considered lucky. Most hotels don't number the thirteenth floor as "13"; they label it "14." Otherwise, people just won't stay in those rooms. The Boeing 707 would never have gotten off the ground if it was called the "Boeing 1300" or "the Boeing 1313." Probably the "Boeing 666" would not have been very popular either.

The Boeing people were smart engineers in their basic design and evolutionary improvements, and they were also smart in naming the beast with a nice numerical scheme that appeals to the numerologists in us. There is, of course, a Boeing 777 now, and that sounds like another improvement—it has an extra 7!

# PART VII

# Do

"Do. Or do not.

There is no try."

Jedi Master Yoda
*Star Wars*

# 43

## Learn by Doing

There is a wonderful story in David Bayles and Ted Orland's *Art and Fear* about learning by doing. An art instructor tells his pottery class that the left side of the classroom will be graded on the total weight of the pots they create during the semester. At the end of the course, the teacher said he'd bring in his bathroom scales and weigh their pots: fifty pounds of pots would be an "A," forty pounds a "B," thirty pounds a "C," and so forth. The right-hand side of the class would be graded on the quality of only one pot. Their job was to make the best pot they could and to turn it in for a judgment on quality alone.

So at the end of the semester, guess what happened. The quantity students not only made the most pots—they also made the best pots. While the quality students sat around and theorized about the perfect pot, the quantity students were busy making lots of pots. The quantity students learned from their mistakes and didn't get hung up on perfection. Their quality steadily improved with the pots they made and they ended up surpassing the quality students.

This "Parable of the Pots" is a story I tell my senior students in their spacecraft design course. I want them to overcome their fears of making mistakes and to learn by doing. (Doers do it better.) Aerospace design has similarities to art. There is no theory of design that works in all cases. (There are many handbooks that have the word "design" in the title, but these books are usually about technique and prior designs.) Creativity is required; something new and interesting should be produced.

Aerospace students are taught a lot of math, physics, and engineering before they are asked to design a spacecraft. These courses are but the background, just as art has a theory of color and perspective. Putting all the techniques together to create a new design is a different story.

Students have been taught piecemeal courses by reductionist methods (i.e., little pictures). Design requires a holistic view—the big picture. It is a small wonder that students approach their senior design course in a state of trepidation. Will what they do add up to anything?

So I tell them the Parable of the Pots to allay their fears. When it comes to design, you learn by doing.

# 44

## Sharpen Your Axe

Two lumberjacks were chopping wood. The first was a burly bear of a man with thick forearms. The second was a wiry, thin man. They were each cutting several cords of firewood for the winter.

The big man wielded his axe with powerful blows. He kept swinging with unrelenting determination and barely took time to wipe the sweat from his brow. The wiry man was also cutting very quickly but would stop every once in a while and leave.

The big man wondered where he went but didn't let this distraction slow him down. He wasn't a quitter.

The wiry man would come back and resume chopping with great gusto. Sometimes their blows on the wood would fall into synchrony. Then they'd chop faster—as if they were in a race. After a few hours the wiry man announced that he was finished. He had completed his task.

The burly man said, "How the hell is that possible? I'm only half done—and you've been taking all those breaks!"

The wiry man said, "I haven't been taking any breaks—I've been sharpening my axe."

The lesson is obvious. When you tackle a difficult problem, make sure you not only start with the best tools but also keep improving your tools as you go along.

Rocket scientists depend on computer tools to solve the problems of spacecraft guidance and control. Very often the managers of these rocket scientists push them to work long hours to finish a task. A contract has come in, a new spacecraft problem has been discovered, a question from upper management has been asked. The question is not a request, it is an order: an action item has been issued and answers are needed in a hurry. "We need the answer yesterday!"

In this kind of environment (which is unfortunately pervasive), there is no time for software improvements. Old programming codes written decades ago (in the not quite dead language of Fortran) are used to solve the new problems—even problems that the programs were not originally designed to solve.

Rocket scientists and anyone who works on difficult problems need time to develop the right tools for the right problem. They need time to think about what they are doing and how to do it best.

If significant blocks of time were allocated to tool development and improvement, the aerospace industry would become more efficient. A great deal of money could be saved and better products could be produced. Our commercial aviation, our military, and our space programs would all benefit.

Rocket scientists need time to sharpen their axes. So do you.

# 45

## Correct It on the Way

In missions to other planets, it was learned that there will always be errors.

In the first U.S. planetary mission, which went to Venus, scientists and engineers argued whether a tank of propellant and a rocket engine would be needed for a midcourse correction maneuver. At first it was believed that we could launch the *Mariner* spacecraft directly at Venus and that the craft would coast all the way to the planet without significant error.

It was like shooting an arrow at a target. Once the arrow leaves the bow you have no control over where it lands. The only control you have is the direction you point it in.

Some engineers calculated the effect of the pointing error of the launch vehicle (and other effects such as the very small force of solar radiation on the spacecraft) and discovered that the *Mariner* spacecraft could easily miss Venus by a million miles. But to perform the scientific observations, the craft had to be within 20,000 miles of the cloud-enshrouded planet. Putting an extra propulsion system on a spacecraft increases the weight and complexity—which increases the cost significantly. Up to this point, rocket scientists had not conceived the idea of a trajectory correction maneuver (or TCM). Eventually, the engineers convinced project managers that a midcourse correction maneuver (and the requisite propulsion system) was absolutely necessary if the *Mariner* spacecraft were to get to Venus.

Nowadays, TCMs are old hat. You wouldn't begin to think of designing an interplanetary mission without incorporating a series of trajectory correction maneuvers to bring the spacecraft back on course. We often hear of the extraordinary accuracy of space shots. "It was like hitting a dime in New York with a rifle in Los Angeles." These analogies are not entirely true. A better analogy was given by my friend Bob Cesarone, who was for all practical purposes the remote pilot of the *Voyager* spacecraft that went to Neptune. The way Bob put it, "It's not like hitting a hole in one. It's more like a series of precise, short putts on a golf course. Each putt gets you closer to the target."

So rocket scientists are not perfect and neither are their spacecraft. Errors are expected and accounted for. Small mistakes will always be present.

By knowing you live in a world of errors, you can plan to take corrective action. Don't be afraid to make trajectory corrections in your life. If you expect they will be necessary, then making a correction will not be an admission of failure but a reflection of wisdom and foresight.

# 46

## Do Something

Two mission designers—one a mathematical type, the other a visualizer—were designing complex trajectory scenarios for a future interplanetary space probe.

The math type knew the equations of motion and the theories of celestial mechanics; he had done a Ph.D. thesis on the subject. But in spite of his wealth of knowledge, or perhaps because of it, he sat at his computer terminal, paralyzed. Many thoughts ran through his head. He analyzed what would happen for a number of ideas as they occurred to him. He was like a chess player in deep thought, thinking of moves, imagining the problems with each move, and then rejecting each move one by one. "If I try this," he said to himself, "then it won't work because blah, blah, blah . . . But if I do that it will lead to this other problem . . . I could try this other idea, but then I paint myself into a corner because of this. . . ."

He sat frozen in front of his keyboard. He'd type a few keystrokes into his trajectory simulator, which would calculate potential orbits for the spacecraft. But then he'd hit the backspace key and stop. He had great difficulties designing a good trajectory for the future space mission.

The other designer rolled up his sleeves and tried things. He didn't worry too much about the results; he just punched away at his keyboard and let the trajectory simulator fly. He made plenty of mistakes, but he also made a lot of progress. He was a fast typist, but more important, he was a keen observer of the results and patterns that showed up on his computer screen. He designed several excellent mission scenarios that the project managers liked.

The first designer had to learn, painfully, to give up his mathematical perfectionism and to try new things; to proceed without a theory, to learn from experience. He had to learn that he was dealing with a complex system from which new properties could emerge. These properties could not easily be predicted from the theories of celestial mechanics that he had studied in his doctoral thesis.

Recently, a new field called complexity theory has begun to provide insights into nonlinear systems and how order can emerge from chaos. The second designer intuitively knew to look for such patterns. He followed the thinking style of the Synthesist, as described by Harrison and Bramson.

In this case, the Synthesist made a better rocket scientist than the Analyst (the first designer). The Synthesist didn't just sit there—he did something.

# 47

## Don't Ignore Trends

There is a tendency to ignore small problems. The gas gauge is low—I still have enough—it can wait. Forget about it.

Then we run out of gas, and it's a catastrophe.

Rocket scientists cannot afford to ignore small problems on a spacecraft in flight. Something must be done to understand the problem before it gets worse. A capacitor on the *Voyager* spacecraft became the most studied electronic component in history because it began to fluctuate (in capacitance), which changed the radio frequency the spacecraft was listening to. So the transmission signal from Earth had to be constantly adjusted so that *Voyager* would hear its commands. It was as if a radio station you were listening to kept drifting so that you would have to keep turning the knob on your receiver to hear it.

The capacitor on the *Voyager* was affected by variations in the onboard temperature. If *Voyager* ever stopped listening to the mission controllers on Earth, it would mean the loss of the spacecraft and the mission. So a great deal of effort went in to analyzing the trends of this faulty capacitor on *Voyager*'s receiver.

In life we are constantly confronted with problems. We are very busy just following our usual schedules, and we don't have time to deal with new problems. When the great figure skater Scott Hamilton was diagnosed with testicular cancer, he was on tour with his Champions on Ice. His first reaction was, "I don't have time for this!" Fortunately, Scott took the time—and his life and career were saved.

We cannot afford to ignore the early warning signs of problems, whether they are a dripping faucet, a low tire, or a seemingly minor health complaint. Scott Hamilton's cancer started out with stiffness and stomach cramps, which he first thought he could work off.

It is not generally known that our deep-space probes need the constant attention of technicians on Earth. These mission controllers are affectionately referred to as "babysitters." Eventually, spacecraft will become fully automated adults, but today (in 2006) they are still babies in the crib. They simply are not smart enough to take care of themselves.

The babysitters continuously monitor the health of the spacecraft. They listen to the spacecraft every day and maintain a constant communication link. The early versions of our interplanetary probes were more like toasters than automated robots. When you toast bread, you push the slice down and set the dial anywhere from light to dark. But the toaster doesn't know what's happening to the bread. It simply times the heating element and pops the toast out. If the toast should burst into flames in the first five seconds, the toaster will continue to heat the element until the timer runs out (say in thirty seconds). This type of control is what rocket scientists call open-loop control, which means no control—just timing without knowing what is really happening.

Spacecraft even today do not have complete feedback control. That is, they don't know what they're doing. A feedback control on a toaster would mean that the toaster would be measuring the temperature of the toast so that once it got hot enough for long enough, it would turn off. If for some reason it burst into flames, it would immediately shut down the heating element, unlike the open-loop toasters. The thermostat control in your home is an example of such a feedback control system.

Spacecraft are still pretty dumb. When we command a spacecraft to take a picture of Saturn's rings (which the *Cassini* spacecraft is doing right now, at the time of this writing), it points its camera, blindly, to a predetermined direction and takes a picture. It does not have the capacity to look through the camera and check if Saturn is in the frame. But the pictures of Saturn's rings came out fine because the mission controllers had been monitoring the trends in the spacecraft for years and correcting for small errors along the way.

Paying attention to trends is an important concept, whether you're babysitting a spacecraft, watching your health, or just burning toast.

# 48.

# Work on Your Average Performance

Rocket scientists have a term they use to describe the average expectation for a space mission. They say everything is "nominal" when things are going as planned, when the spacecraft is following the average trajectory out of thousands they simulated in their computers before flying the real mission.

All those simulations are like the practicing that athletes, musicians, and actors do before the actual performance. There is a big difference between knowing (the theory of) how to do something and the actual performing. Parents say that their toddler understands a lot, even though she doesn't talk yet. This is because understanding speech is easier than speaking. Listening is understanding language, but performing is putting language into practice.

A coach was asked how his world-class figure skater would do in a competition for world champion. He said, "If she manages to do her average performance, she will win, because no one can beat her average performance." This is why people say you should prepare for the worst and hope for the best.

Spaceflight is a performance. Years of planning, simulation, and practice are behind it. Rocket scientists spend a great deal of time simulating what could go wrong and how to fix it when it does. They worry about Murphy's law. They know something will go wrong—that's why all good designs have plenty of backup systems. But in the end, in the real mission, usually something in between the extremes happens. The worst-case scenario doesn't occur; neither does the best.

The spacecraft does its nominal performance. The rocket scientists are very happy to hit their average—it is a very high standard, which is why spacecraft sometimes last a lot longer than the original planned mission. Often, rocket scientists are just a little lucky and the spacecraft does better than the nominal mission.

The *Voyager 1* and *2* spacecraft were planned to encounter Jupiter and Saturn with a 95 percent probability of success. If all went well with *Voyager 1*, then *Voyager 2* would be targeted to Uranus with a 60 percent chance of success and to Neptune with only a 40 percent chance of success. So the scientists didn't really expect that *Voyager* would make it all the way to Neptune. It was a bit of a long shot.

In fact, both spacecraft had problems shortly after launch. The rocket scientists had tried to make the twin *Voyagers* smart (by adding fault-protection software), but then the spacecraft started acting funny (refusing to obey commands because the spacecraft got too twitchy about taking risks) and so their programming had to be ripped out and replaced (all this during flight) to make the spacecraft obey direct orders from Mission Control. A few times the babysitters, who constantly monitored the health of the spacecraft, lost contact, and the engineers had to do some detective work to figure out what went wrong and how to reestablish the communication link.

One time, one of the *Voyagers* locked on the wrong star (a case of mistaken identity) and pointed its antenna not toward Earth but in an entirely different direction—where it thought Earth was.

Fortunately *Voyager 2* had a great mission operations team, and the spacecraft eventually made the whole trip from Jupiter to Saturn to Uranus and to Neptune—traveling a distance of nearly 4 billion miles from the Earth in twelve years. It was one of the most successful missions in the history of space exploration.

*Voyager 2* performed just a little better than its average.

# 49

## Look Behind You

Gramps (my Dad's father) never looked through his rearview mirror when he drove. When asked about this he'd say, "What do I care about where I've been?"

But rocket scientists have to look backward and forward when they are conducting a space mission. During spaceflight, they have to determine precisely where the spacecraft is. This is a very difficult task because they cannot see the spacecraft and there is no Global Positioning System (like we have around Earth for pinpoint terrestrial navigation) in the solar system. Rocket scientists need the spacecraft's precise location and velocity so they can predict where it will be in the future—how close it will come to its destination (a planet, an asteroid, or a comet). A tiny error in this determination can mushroom into a very large error down the road—causing the spacecraft to miss its target by millions of miles. (This problem is closely related to that of navigating a ship on the open seas, where a small error in the ship's chronometer could result in missing an island by hundreds of miles or crashing on a reef; for a brilliant account, see Dava Sobel's national bestseller, *Longitude*.)

To reduce the error, rocket scientists must know where the spacecraft has been in the past through a process called trajectory reconstruction. It comes down to the simple concept that, "If you don't know where you were or when you were there, it's going to be difficult to go where you want to go."

Which reminds me of a stupid joke I heard at JPL about engineers and mathematicians. (We had lots of them—engineers and mathematicians, and stupid jokes.) Managers at the lab would complain that, "The mathematicians will tell you something that is absolutely true (and they can prove it) but it's absolutely useless; the engineers will give you precise numbers (which they obtained from accurate measurements) that mean precisely nothing." The crux of the joke turns on the cultural difference between the two disciplines. Ready? Two engineers were stranded in a lifeboat on the ocean. A balloon with two mathematicians drifted overhead. The engineers shouted up at the mathematicians, "Can you tell us for certain where we are?" The mathematicians yelled down, "Yes, absolutely—but first tell us our precise location." Then the engineers, after a while, shouted up at the mathematicians, "You are in a balloon precisely ten meters above us. It is a number you can rely on." Then the mathematicians yelled down at the engineers, "You are in the Pacific Ocean. There can be no doubt of it."

See—I told you it was stupid.

But jokes like these remind rocket scientists about their dilemma in space navigation—it is difficult to get precise information about where you are because everything is relative. The spacecraft may be close to a planet, but where is the planet? We don't have precise knowledge of where the planets are located.

A lesson in life can be drawn. Sometimes it helps to reflect on where we've been when we make decisions about our future. Who are we? What do our lives stand for? Where are we going? We get meaning and guidance from the background of our lives. A wise person said, "Every day I become more the person I was meant to become." Socrates said, "The unexamined life is not worth living."

It bothered a lot of people that Gramps never looked back (especially the drivers behind him). He focused only on getting to where he was going, not on where he had been. Fortunately (somehow), he never had a serious accident and lived to the ripe old age of 91. But his driving habit left a lot to be desired.

Not looking behind you can be as dangerous as not looking around you or ahead of you.

Ask any insurance agent.

# 50

## Learn from Your Mistakes

"I do experiments to be embarrassed," said the young professor of rocket science.

He had a very interesting and humble way of looking at the reason he did laboratory work. Before doing an experiment, you must have a theory that you are testing. It doesn't have to be your own theory, although it might be. It could be Einstein's general theory of relativity that you are putting to the test. According to Einstein's theory, a specific outcome would be expected, so we say that Einstein predicts a certain result.

Sometimes people get the wrong impression of science and think it is a belief system; that scientists (and everyone else) have to believe Einstein because he was a great genius, so whatever he said must be true and so we set up experiments to find more evidence that he was right. But science is the opposite of belief because scientists are often trying to prove that a previous theory is wrong. When a test of relativity is done, the scientist who does it is challenging Einstein—the experiment could prove that Einstein was wrong. The scientist who does the experiment could achieve a kind of immortality for toppling a great genius. Because Einstein's theory is so well established, proving him wrong would be worth a Nobel prize, even if the scientist who did it didn't have a new theory. This is how scientists become famous. Einstein did it to Newton by proving that Newton's law of gravity could not explain the motion of the planet Mercury, whereas Einstein's theory explained the motion precisely and also predicted the bending of starlight by the sun, which was later proved by photographs of a solar eclipse.

So that's why the young professor wanted to be embarrassed, so he could learn something new about the universe; getting embarrassed is tantamount to becoming famous. Maybe he wasn't that humble.

When mistakes are made in the space program, the results can be far worse than mere embarrassment. Rockets explode, billion-dollar satellites are destroyed, lives are lost. Sometimes the mistakes are due to poor workmanship, bad design, insufficient oversight, or lack of funding. In the best cases, when everything is done right, we can still be surprised.

We go into space to learn new things. Space is the most hostile of all environments, but it also offers us a wealth of power and a source of security for the future of humanity. The average American supports NASA's space program and is very forgiving of the agency's mistakes, because most people understand that space travel is extremely hazardous and that the risks are great.

But not all of NASA's mistakes are innocent lessons that come with the territory. Some mistakes were mistakes on the drawing board. NASA misled Congress about the safety, reliability, and economy of the shuttle. Richard Feynman makes the case in his penetrating book *What Do You Care What Other People Think?* It is time for NASA to go back to the great lessons that were learned in getting to the moon. It is time for NASA to abandon the shuttle, to design unmanned heavy-lift inexpensive boosters, and smaller human-rated launchers. It is time for NASA to set their goal on human exploration of Mars with a comprehensive plan for getting there.

It is time for NASA to start thinking like rocket scientists again.

# Epilogue

There are mistakes in this book: typos, grammatical goofs, factual faux pas.

How do I know this?

Because of my experience with the rocket scientist's nemesis: Murphy's law. I, of course, take full responsibility. Better yet, I will take your "trajectory corrections" and put them in a future edition.

Was I too hard on NASA?

Let me know.

My greatest dream is to see people on Mars, not just within my lifetime, but soon. Let's not just think about it—let's do it. Recently (in September 2005), NASA's new administrator, Dr. Michael Griffin, declared that the space shuttle, the international space station, and nearly the entire U.S. manned space program for the past three decades were mistakes. It appears that Dr. Griffin will be putting NASA back on track and soon they will be thinking like rocket scientists again!

I hope I have demystified the thinking of a rocket scientist. Rocket science is just common sense applied to the extraordinarily uncommon environment of outer space. (And rocket scientists are people, too!)

Please let me know what you think. With your permission, I may use your corrections (typos, etc.) and contributions (ideas and anecdotes) in a future edition of *How to Think Like a Rocket Scientist* (and I will acknowledge your help).

# Recommended Viewing: The Greatest Sci-Fi Films of the Twentieth Century

In rank order, here's my list.

## 1. *The Day the Earth Stood Still* (1951)

This is a film that has stood the test of time because it concentrates not on alien special effects but on human reactions, fears, and failings in the light of a cosmic visitation. The effects are simple, pure, and still believable. When the robot's visor begins to open, you will remember the words "Gort, Klaatu barada nikto!" which prevent your vaporization. Professor Barnhard's blackboard equations are real, celestial mechanics equations, and his discussion with Klaatu, the alien, is technically accurate. The characters are wonderful, and the message is as relevant today as it was in 1951.

What more can we ask of a sci-fi film?

## 2. *Forbidden Planet* (1956)

Robby the Robot is a lot friendlier than Gort—he's incapable of harming human beings. He was tinkered together by Dr. Morbius (Walter Pidgeon) who is the sole survivor (except for his beautiful daughter) on a planet that once was the home of the Krell—a civilization a million years ahead of Earth's. A rescue party wants to take Dr. Morbius and his daughter back to Earth, but Morbius refuses and warns the captain that they are in grave danger. Dr. Morbius is obsessed with studying the Krell and learning of their demise. The next night a crew member is murdered. The ship's doctor slips into Morbius's lab to take the IQ boost to find out what's going on. The jolt fatally injures him, but before he dies he reveals all to the captain. When Morbius sees the doctor's body he exclaims, "The fool—to think that his ape-brain could contain the secrets of the Krell!" In the end, we learn what the Krell's final project was, how they were all killed in a single night, and what's killing the crew of the rescue ship.

### 3. *Blade Runner* (1982)

Skip the director's cut. Ridley Scott made a big mistake by eliminating the voice-over, which reveals what the blade runner (Harrison Ford) is thinking. As we listen to the main character's thoughts, we realize that he's not the brightest porch light on the block nor the most moral cop. He's a gun for hire, and his current assignment is to retire (kill) six slave androids who have returned to Earth after murdering dozens of people. The most interesting character is Roy (Rutger Hauer), who wants his creator (Joe Turkel) to extend his life beyond the four-year termination date. The film ends with a powerful scene of transformation and enlightenment. Even the blade runner gleans a glimmer of understanding.

### 4. *Dune* (1984)

This is a rich and complex film (based on the book by Frank Herbert) that rewards you with every viewing. The year is 10,991 (A.D. presumably) and the human universe consists of many planets light-years apart. There is a back history of a human revolt over conscious machines and robots (when people were enslaved by them) so that special guilds were founded to develop human powers. Certain women have the power of the voice, of truth saying, and of telepathy. Some men have savant skills of supreme logic and mathematical calculation. A feudal system of houses governs each planet, and an emperor rules over the universe. A plan unfolds to overthrow the emperor, and it is foretold that a superbeing (the product of a secret breeding program) will emerge to replace him.

### 5. *2001: A Space Odyssey* (1968)

Stanley Kubrick's masterpiece set the special-effects standard for all future sci-fi films. This is the story of alien contact with human beings told in biblical style. The first scene is an interpretation of Genesis: The Dawn of Man. We are not told what the black monolith is. Kubrick does not expect us to understand our first contact with advanced alien intelligence. In the year 2001 (a year Kubrick never saw, as he died in 2000), humans will have a shuttle, a space station, a lunar ferry, and a colony on the moon. An interesting subplot is that we'll have computers that are more conscious and personable than our astronauts. And more treacherous. The music is classical and just as rich as the visual effects. Eventually, one astronaut rides the wormhole to the alien planet. The one defect of the film is the length of the ride (but you can always fast-forward). The film ends with transformation and no explanation. Enjoy the mystery—you aren't expected to understand it entirely.

### 6. *Dr. Strangelove, or: How I Learned to Stop Worrying and Love the Bomb* (1964)

Another classic by Stanley Kubrick, who decided that thermonuclear combat was too depressing to be filmed as a serious drama—so he made it a satire. It's amazing how slight a shift he made to slip from cold reality to comic consciousness. The film features Peter Sellers in three roles (his greatest performances), brilliant comedy from the great natural actor, George C. Scott, who plays General Turgidson, and the rodeo clown antics of Slim Pickens who, as Major Kong, proves that American stick-to-it-iveness can deliver the bomb no matter what the Ruskies throw at us. A must see for every high school student—you shouldn't get your diploma without having viewed it.

### 7. *The Fly* (1958)

The ending is so tragic that they put it first, so we can get over it. A scientist is dead and his wife is a suspect. The rest of the movie is the wife's story of the idyllic life she shared with her husband, a brilliant scientist in the process of creating his greatest invention: the disintegration–integration machine, or in *Star Trek* parlance, the transporter. The movie features a beautiful romance, a who-dunit mystery, and a sci-fi horror story, with great dialogue, excellent acting, and an impassioned eulogy from the scientist's brother (Vincent Price) who compares his sibling to a pioneer in a dangerous wilderness.

# Recommended Reading
# and Bibliography

Aldrin, Buzz and McConnell, Malcolm, *Men from Earth*, Bantam Books, New York, NY, 1989.

Asimov, Isaac, *I. Asimov: A Memoir*, Bantam Books, New York, NY, 1995.

Auden, W. H. and Kronenberger, Louis, *The Viking Book of Aphorisms: A Personal Selection*, Barnes & Noble Books, New York, NY, 1993.

Augustine, Norman R., *Augustine's Laws: And Major System Development Programs*, American Institute of Aeronautics and Astronautics, Inc., New York, NY, 1983.

Bartlett, John, *Bartlett's Familiar Quotations*, Ed., Kaplan, Justin, 16th Edition, Little, Brown and Company, Boston, MA, 1992.

Bayles, David and Orland, Ted, *Art and Fear: Observations on the Perils and Rewards of Artmaking*, Image Continuum Press, Santa Cruz, CA, 2001.

Bell, E. T., *Men of Mathematics*, Simon and Schuster, New York, NY, 1965.

Burrows, William E., *Exploring Space: Voyages in the Solar System and Beyond*, Random House, Inc., New York, NY, 1990.

Burrows, William E., *This New Ocean: The Story of the First Space Age*, Random House, Inc., New York, NY, 1999.

Clarke, Arthur C., *2001: A Space Odyssey*, New American Library, New York, NY, 1968.

Clary, David A., *Rocket Man: Robert H. Goddard and the Birth of the Space Age*, Hyperion, New York, NY, 2003.

Davidson, Keay, *Carl Sagan: A Life*, John Wiley & Sons, Inc., New York, NY, 1999.

Dewdney, A. K., "Word ladders and a tower of Babel lead to computational height defying assault," *Scientific American*, Aug. 1987, pp. 108–111.

Doyle, Arthur Conan, *The Complete Sherlock Holmes*, CRW Publishing Limited, London, England, 2005.

Easterbrook, Gregg, "Big Dumb Rockets," *Newsweek*, Aug. 17, 1987, pp. 46–60.

Feynman, Richard P., *"Surely You're Joking, Mr. Feynman!" Adventures of a Curious Character*, Bantam Books, Inc., New York, NY, 1988.

Feynman, Richard P., *"What Do You Care What Other People Think?" Further Adventures of a Curious Character*, W. W. Norton & Company, New York, NY, 2001.

Flesch, Rudolf, *The Art of Clear Thinking*, Barnes & Noble Books, New York, NY, 1973.

Fox, Jeffrey J., *How to Become a Great Boss: The Rules for Getting and Keeping the Best Employees*, Hyperion, New York, NY, 2002.

Gehman, Jr., Harold W., et al., *Columbia Accident Investigation Board Report*, Vol. 1, NASA, Washington, DC, 2003.

Gelb, Michael J., *How to Think Like Leonardo da Vinci: Seven Steps to Genius Every Day*, Dell Publishing, New York, NY, 2000.

Gelb, Michael J., *Discover Your Genius: How to Think Like History's Ten Most Revolutionary Minds*, HarperCollins Publishers, Inc., New York, NY, 2003.

Gerrold, David, *The World of Star Trek*, Ballantine Books, New York, NY, 1974.

Gladwell, Malcolm, *The Tipping Point: How Little Things Can Make a Big Difference*, Back Bay Books, Boston, MA, 2002.

Harris, Allen R. and Bramson, Ph.D., Robert M., *The Art of Thinking: Strategies for Asking Questions, Making Decisions, and Solving Problems*, The Berkley Publishing Group, New York, NY, 1984.

Harris, Judith Rich, *The Nurture Assumption: Why Children Turn Out the Way They Do*, Touchstone, New York, NY, 1999.

Hart, Michael H., *The 100: A Ranking of the Most Influential Persons in History*, Citadel Press, New York, NY, 1994.

Hirsch, Jr., E. D., Kett, Joseph F., and Trefil, James, *The Dictionary of Cultural Literacy: What Every American Needs to Know*, Houghton Mifflin Company, Boston, MA, 1993.

Hofstadter, Douglas R., *Gödel, Escher, Bach: An Eternal Golden Braid*, Vintage Books, New York, NY, 1980.

Kraft, Chris, *Flight: My Life in Mission Control*, Plume, New York, NY, 2002.

Longuski, Jim, *Advice To Rocket Scientists: A Career Survival Guide for Scientists and Engineers*, American Institute of Aeronautics and Astronautics, Reston, VA, 2004.

Lovell, Jim and Kluger, Jeffrey, *Apollo 13*, Pocket Books, New York, NY, 1995.

Mencken, H. L., *A New Dictionary of Quotations on Historical Principles from Ancient and Modern Sources*, Alfred A. Knopf, Inc., New York, NY, 2001.

Morrell, Margot and Capparell, Stephanie, *Shackleton's Way: Leadership Lessons from the Great Antarctic Explorer*, Viking Penguin, New York, NY, 2001.

Mullane, Mike, *Riding Rockets: The Outrageous Tales of a Space Shuttle Astronaut*, Scribner, New York, NY, 2006.

Nimoy, Leonard, *I Am Not Spock*, Ballantine Books, New York, NY, 1977.

Nimoy, Leonard, *I Am Spock*, Hyperion, New York, NY, 1995.

Parker, Barry, *Einstein: The Passions of a Scientist*, Prometheus Books, Amherst, NY, 2003.

Parker, Barry, *Albert Einstein's Vision: Remarkable Discoveries That Shaped Modern Science*, Prometheus Books, Amherst, NY, 2004.

Pirsig, Robert M., *Zen and the Art of Motorcycle Maintenance*, Bantam Books, Inc., New York, NY, 1976.

Poundstone, William, *Carl Sagan: A Life in the Cosmos*, Henry Holt and Company, New York, NY, 1999.

Sacks, Oliver, *The Man Who Mistook His Wife for a Hat and Other Clinical Tales*, Harper Perennial, New York, NY, 1990.

Schow, David J., *The Outer Limits Companion*, GNP/Crescendo Record Co., Inc., Publishing Division, Hollywood, CA, 1998.

Schumacher, E. F., *Small Is Beautiful: Economics as if People Mattered*, Harper & Row, Publishers, Inc., New York, NY, 1975.

Schwartz, Carol, *Video Hound's Sci-Fi Experience: Your Quantum Guide to the Video Universe*, Visible Ink Press, Detroit, MI, 1997.

Scott, William B., "Systems Strategy Needed To Build Next Aero Workforce," *Aviation Week & Space Technology*, May 6, 2002, pp. 61–62.

Sobel, Dava, *Longitude: The True Story of a Lone Genius Who Solved the Greatest Scientific Problem of His Time*, Penguin Books, New York, NY, 1995.

St. James, Elaine, *Simplify Your Life: 100 Ways to Slow Down and Enjoy the Things That Really Matter*, Hyperion, New York, NY, 1994.

Stoll, Cliff, *The Cuckoo's Egg: Tracking a Spy Through the Maze of Computer Espionage*, Pocket Books, New York, NY, 1990.

Stoll, Clifford, *Silicon Snake Oil: Second Thoughts on the Information Highway*, Doubleday, New York, NY, 1995.

Stoll, Clifford, *High-Tech Heretic: Why Computers Don't Belong in the Classroom and Other Reflections by a Computer Contrarian*, Doubleday, New York, NY, 1999.

Thorpe, Scott, *How to Think Like Einstein: Simple Ways to Break the Rules and Discover Your Hidden Genius*, Barnes & Noble Books, New York, NY, 2002.

Wenger, Ph.D., Win and Poe, Richard, *The Einstein Factor: A Proven New Method for Increasing Your Intelligence*, Prima Publishing, Roseville, CA, 1995.

Whitfield, Stephen E. and Roddenberry, Gene, *The Making of Star Trek*, Ballantine Books, New York, NY, 1974.

Wolfe, Tom, *The Right Stuff*, Bantam Books, Inc., New York, NY, 2001.

Zubrin, Robert and Wagner, Richard, *The Case for Mars: The Plan to Settle the Red Planet and Why We Must*, Touchstone, New York, NY, 1997.

# About the Author

After receiving his Ph.D. in aerospace engineering from The University of Michigan in 1979, Jim Longuski (lŏng-gŭs'-skē) worked at the Jet Propulsion Laboratory as a Maneuver Analyst and as a Mission Designer. In 1988, he joined the faculty of the School of Aeronautics & Astronautics at Purdue University in West Lafayette, Indiana, where he teaches courses in dynamics, aerospace optimization, and spacecraft design. He is coinventor of a "Method for Velocity Precision Pointing in Spin-Stabilized Spacecraft or Rockets" and is an associate fellow of the American Institute of Aeronautics and Astronautics (AIAA). Professor Longuski has published more than one hundred fifty conference and journal papers in the general area of astrodynamics including such topics as spacecraft dynamics and control, reentry theory, mission design, space trajectory optimization, and a new test of general relativity. In 2004, AIAA published Professor Longuski's first book, *Advice to Rocket Scientists: A Career Survival Guide for Scientists and Engineers.*

# About the Illustrator

Masataka Okutsu was born in Yamanashi-ken, Japan. He obtained his B.S. (in 1999) and M.S. (in 2001) in aeronautical and astronautical engineering from Purdue University. He is currently working on his doctoral thesis on "Robotic and Human Space Missions" under the direction of Professor James M. Longuski.

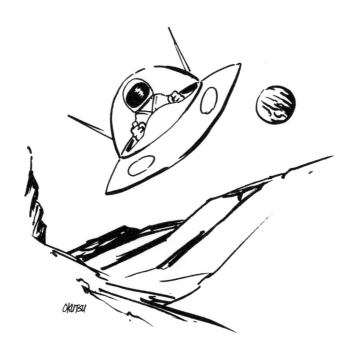

# Index

Printed in the United States of America.